# Springer Theses

Recognizing Outstanding Ph.D. Research

## Aims and Scope

The series "Springer Theses" brings together a selection of the very best Ph.D. theses from around the world and across the physical sciences. Nominated and endorsed by two recognized specialists, each published volume has been selected for its scientific excellence and the high impact of its contents for the pertinent field of research. For greater accessibility to non-specialists, the published versions include an extended introduction, as well as a foreword by the student's supervisor explaining the special relevance of the work for the field. As a whole, the series will provide a valuable resource both for newcomers to the research fields described, and for other scientists seeking detailed background information on special questions. Finally, it provides an accredited documentation of the valuable contributions made by today's younger generation of scientists.

## Theses are accepted into the series by invited nomination only and must fulfill all of the following criteria

- They must be written in good English.
- The topic should fall within the confines of Chemistry, Physics, Earth Sciences, Engineering and related interdisciplinary fields such as Materials, Nanoscience, Chemical Engineering, Complex Systems and Biophysics.
- The work reported in the thesis must represent a significant scientific advance.
- If the thesis includes previously published material, permission to reproduce this must be gained from the respective copyright holder.
- They must have been examined and passed during the 12 months prior to nomination.
- Each thesis should include a foreword by the supervisor outlining the significance of its content.
- The theses should have a clearly defined structure including an introduction accessible to scientists not expert in that particular field.

More information about this series at http://www.springer.com/series/8790

Dimitry Mikitchuk

# Investigation of the Compression of Magnetized Plasma and Magnetic Flux

Doctoral Thesis accepted by
the Weizmann Institute of Science, Rehovot,
Israel

 Springer

*Author*
Dr. Dimitry Mikitchuk
Department of Physics of Complex Systems
Weizmann Institute of Science
Rehovot, Israel

*Supervisor*
Prof. Yitzhak Maron
Faculty of Physics
Weizmann Institute of Science
Rehovot, Israel

ISSN 2190-5053 ISSN 2190-5061 (electronic)
Springer Theses
ISBN 978-3-030-20857-8 ISBN 978-3-030-20855-4 (eBook)
https://doi.org/10.1007/978-3-030-20855-4

This Springer imprint is published by the registered company Springer Nature Switzerland AG
The registered company address is: Gewerbestrasse 11, 6330 Cham, Switzerland

# Supervisor's Foreword

It is my pleasure to introduce Dr. Dimitry Mikitchuk's Ph.D. research for publication in the *Springer Thesis* series. Dr. Mikitchuk was awarded his Ph.D. from the Weizmann Institute of Science in January 2017 for the research presented in this book. The main subject of his experimental study is the investigation of the compression of magnetized plasma and magnetic field by plasma implosion. This subject is relevant to the Magnetized Liner Inertial Fusion and likely to astrophysical plasmas, such as sunspots or other astrophysical objects where the magnetic flux is frozen in an imploding plasma. Here, the magnetized plasma and magnetic flux compression are achieved by using a Z-pinch configuration with preembedded axial magnetic field. A pulsed axial current is driven through the plasma column generating an azimuthal magnetic field that through the Lorentz force compresses the magnetized plasma and the magnetic flux. The diagnostics of the magnetic fields are performed using a noninvasive spectroscopic technique based on the polarization properties of the Zeeman components of different atomic (or ionic) transitions, which enhances the sensitivity of the measurement. In his Ph.D. research, Dr. Mikitchuk made a highly important contribution to the understanding of the physics involved in magnetized plasma compression by the successful determination of the magnetic-field and current-density distributions in the non-equilibrium, transient plasmas. Specifically, his measurements include:

(i) Development and implementation of localized magnetic-field spectroscopic diagnostics for pulsed-power systems based on the polarization properties of the Zeeman effect and using dopant species introduced by laser ablation.

(ii) Direct measurement, for the first time, of the compressed axial magnetic field evolution and distribution during the implosion and stagnation in a Z-pinch with preembedded axial magnetic field utilizing noninvasive spectroscopic methods.

(iii) Simultaneous measurement of the axial and azimuthal magnetic fields revealing unexpected results of the current distribution and the nature of the pressure balance of the axial and azimuthal fields.

(iv) Investigation and demonstration of the mitigating effects of preembedded axial magnetic fields on magneto-Rayleigh-Taylor instabilities in Z-pinch implosions, using interferometric and imaging methods.

These measurements are basic and essential for the advancement of the understanding of complex plasma systems, both in laboratory and in nature. Since the magnetic field is a key factor in magneto-hydrodynamics modeling, the results are highly important for examining simulations, as well as for designing plasma configurations that are particularly relevant to the presently central Magnetized Liner Inertial Fusion approach.

Rehovot, Israel                                                      Prof. Yitzhak Maron
January 2019

# Abstract

In this research, I investigated fundamental phenomena occurring as magnetic-field flux and magnetized plasma are compressed by applied azimuthal magnetic fields. This subject is relevant to numerous studies in laboratory and space plasmas. Recently, it has gained particular interest due to the advances in producing plasmas of high temperature and density in experiments based on the approach of magnetized plasma compression [1]. Many in the plasma physics community consider this approach to be the most promising for achieving controlled nuclear fusion. To advance this approach, it is essential to study experimentally the governing physical mechanisms that take place during the compression. Performing the required systematic experiments is impractical in large-scale facilities designed for fusion demonstration.

In our experiment, we employ a cylindrical (Z-pinch) configuration, in which a current (300 kA, rise time 1.6 $\mu$s) driven through a cylindrical plasma causes implosion of the plasma under the self-generated azimuthal magnetic fields ($B_\theta$). However, our cylindrical plasma is initially embedded in an axial magnetic field $B_z$. The field is quasi-statically applied prior to the high-current discharge, with a value of 0.4 T.

Here, for the first time in these researches, Zeeman-splitting observations are used to measure the evolution and spatial distribution of $B_z$ and $B_\theta$ during the implosion and stagnation stages. The two fields are measured simultaneously, which is rather important due to the irreproducibility that characterizes such experiments of high-current pulses. The difficulties in these measurements are due to (1) the high electron densities in the plasma giving rise to large Stark broadening that smears out the Zeeman pattern, (2) the difficulty in distinguishing between $B_z$ and $B_\theta$, and (3) the absence of light emission from the center of the plasma column. Indeed, in previous studies, under similar conditions, the $B$-fields were only indirectly estimated from the plasma radius. These challenges were achieved by employing a novel spectroscopic technique based on the polarization properties of Zeeman split emission, combined with a laser-generated doping technique that provided mm-scale spatial resolution.

Systematic measurements were performed for different initial conditions of $B_z$ and gas loads. The measurements showed that estimates of the $B$-fields based on the plasma radius are subjected to large errors and thus unreliable. Indeed, the simultaneously measured $B_z$ and $B_\theta$, together with the plasma radius and the discharge current, showed that the application of an initial $B_z$ has a dramatic effect on the current distribution in the plasma. While without $B_z$ the entire current is found, as expected, to flow through the imploding plasma, when an initial $B_z$ is applied, the measured $B_\theta$ (through $\nabla \times \vec{B} = \mu_0 \vec{j}$) showed that only a small part of the current flows within the outer radius of the imploding plasma. Specifically, when $B_{z0} = 0.4$ T, the value of $B_\theta$ in the imploding plasma shell remains nearly constant (between 1.5 and 2 T) during the implosion, even though the current rises and the plasma radius drops. This finding indicates that for implosions with $B_{z0} > 0$ large fraction of the current flows in the peripheral plasma residing at radii larger than the imploding plasma radius. A theoretical model, based on the development of a force-free current configuration in the peripheral plasma, is suggested to explain this unexpected phenomenon. To rigorously test this model, self-consistent 3D MHD simulations are required.

In addition to these results, the measurements provide much information useful for the understanding of the $B_z$-embedded plasma implosion. We measure at stagnation a $\sim 15\times$ compression of the initial axial $B$-field. This compression factor, together with the observed plasma radius, allows for obtaining the $B_z$ confinement efficiency, which is found to be $\sim 50\%$. This information is useful for testing MHD codes. Another phenomenon observed is an axial gradient of $B_z$ in which its magnitude increases by a factor of 2 from the anode (low $B_z$) to the middle of the plasma column ($z \sim 5$ mm, high $B_z$). This measurement demonstrates the existence of a transition region from the uncompressed $B_z = B_z(t = 0)$ inside the electrodes to the compressed $B_z$ farther away from the electrode surface.

The spectroscopic measurements were complemented by 2D images of the plasma self-emission and by interferometric images. These measurements were important both for obtaining the $B$-field evolution and for the study of the dependence of instabilities on the different initial conditions. The measurements clearly showed the mitigation effect of $B_z$ on the magneto-Rayleigh-Taylor instabilities.

The 2D images have also shown the existence of axially directed, filament-like regions that have significantly higher emission than the surrounding plasma. These filaments were found to be plasma regions with higher electron density (by 10–20%), and slightly lower electron temperature (by a few percent) than of the surrounding plasma.

# Reference

1. Gomez MR, Slutz SA, Sefkow AB, Sinars DB, Hahn KD, Hansen SB, Harding EC, Knapp PF, Schmit PF, Jennings CA, Awe TJ, Geissel M, Rovang DC, Chandler GA, Cooper GW, Cuneo ME, Harvey-Thompson AJ, Herrmann MC, Hess MH, Johns O, Lamppa DC, Martin MR, McBride RD, Peterson KJ, Porter JL, Robertson GK, Rochau GA, Ruiz CL, Savage ME, Smith IC, Stygar WA, Vesey RA (2014) Experimental demonstration of fusion-relevant conditions in magnetized liner inertial fusion. Phys Rev Lett 113:155003

# Acknowledgements

I would like to thank people from the Plasma Laboratory of the Weizmann Institute of Science that helped and supported me during the course of my Ph.D. research:

My supervisor Prof. Yitzhak Maron for being a critical and inspiring scientific mentor, for teaching me the art of clear physical thinking, and for setting an example of an uncompromising work.

Dr. Ramy Doron, for his invaluable guidance, patience, expertise, sharp eye, and mind, which have all contributed immensely to the success of this work.

Dr. Eyal Kroupp, for constructive advises during this work, for introducing me to the pulsed-power world, and for being an example of what an experimental physicist should be.

Pesach Meiri for the unexcelled technical support, for his bright solutions for apparently hopeless technical problems, and for the everyday help in any possible experimental complication.

All the technicians in the physics department workshop, and especially to the highly skillful hands of Yehuda Asher.

Dr. Evgeny Stambulchik, Dr. Alexander Starobinets, Dr. Vladimir Bernshtam, Dr. Yuri Kuzminikh, Dr. Subir Biswas, and Dr. Yuri Zarnizki for the collisional-radiative modeling, data analysis, and for productive scientific discussions.

Prof. Amnon Fruchtman and Dr. Henry Strauss for many fruitful discussions on the theoretical side of this research.

Prof. Amnon Fisher for sharing his endless knowledge and experience in any field of the experimental physics.

Dr. Marko Cvejic for being a great laboratory partner with his cheerful and motivating spirit, his hard work, and his interest in our common research.

Last but not least, I would like to thank my family, especially my wife, colleague Christine Stollberg for her moral support, for the possibility to discuss physical problems at the dinner table, and for her help in the MATLAB and LaTeX programming. Furthermore, I want to thank our recently born daughter Yael Stollberg, who gave me the opportunity to work on scientific problems also during numerous sleepless nights.

# Contents

# Chapter 1
# Introduction

## 1.1 Motivation

The evolution of a magnetic field that is embedded in a conducting fluid is closely linked to the fluid motion. Because of the freezing of the magnetic field flux into the conducting fluid, the fluid carries with it the magnetic field lines. Due to the fluid motion, the magnetic field lines are often compressed, bent, and even get distorted, until processes of diffusion allow reconnection of lines and a change of their topology.

Plasmas exhibit this behavior of conductive fluids in space and laboratory, as they have high conductivity and are amenable to motion due to electromagnetic forces. There are ample examples in space plasmas for this coupled plasma motion and magnetic field evolution. The solar wind makes the earth magnetosphere asymmetrical by impinging on it, pushing on the magnetopause at the sun side and entraining the magnetic field lines forming the elongated magnetotail on the night side [1]. On the photosphere of the sun the intensity of the magnetic field reaches kG level in small regions, apparently due to compression of flux tubes by plasma flow [2].

Plasma and magnetic-field flux compression is an important process also in magnetized plasmas in the laboratory. In certain magnetized plasmas, the plasma motion is manipulated by the magnetic field, such as in pinches [3], Tokamaks [4], and plasma thrusters [5]. In other magnetized plasmas the magnetic field evolution is controlled by the plasma motion [6].

Recently, the phenomenon of simultaneous compression of the magnetic field and plasma has steered much interest because of its potential application for controlled thermonuclear fusion [7]. Pulse-power-driven Z-pinches are being considered as sources of x-rays upon stagnation [8]. These x-rays could be used to drive inertial confinement fusion (ICF) capsules. However, the efficiency of converting the kinetic energy of the pinch into x-rays and then back to kinetic energy of the imploding target-capsule is very low. The alternative path for fusion within ICF, proposed in the current configuration by Slutz et al. [7] in 2010, is to first magnetize the fusion fuel within a liner by external field coils. An axial current is then applied and compresses the

© Springer Nature Switzerland AG 2019
D. Mikitchuk, *Investigation of the Compression of Magnetized Plasma and Magnetic Flux*, Springer Theses,
https://doi.org/10.1007/978-3-030-20855-4_1

solid-cylinder and the fuel inside it. The initial axial magnetic field is expected to be frozen into the fuel and subsequently to be compressed as the liner implodes. Ideally, due to the implosion the magnetic field strength is expected to rise dramatically and confine the fuel-plasma in the radial direction, thereby enabling the formation of the desirable conditions for fusion. For the realization of this attractive scenario it is crucial to understand the governing physical mechanisms. For example, one has to examine to what extent the magnetic field is frozen into the plasma and whether the plasma resistivity allows for diffusion of the magnetic field, which competes with the flux compression.

Laboratory experiments, in which the coupled evolution of magnetic field and plasma is explored, can greatly contribute to the understanding of these fundamental processes (e.g., [9–11]). The recent rising interest in compressing magnetized plasma follows more than two decades of related experimental [12–17] and theoretical [18, 19] research. However, reliable detailed diagnostics, particularly that of the magnetic field evolution, which is essential to advance the research, is still missing. Therefore, the main goal of the present study was detailed investigation of the magnetic field evolution and distribution in a magnetized plasmas. Here, a laboratory experiment on flux compression is studied by the application of advanced, non-intrusive, spectroscopic techniques to measure the plasma conditions and the magnetic field. The investigated configuration consists of a cylindrical gas-puff from a nozzle with an embedded (initially uniform) axial field ($\vec{B}_z = B_{z0}\hat{z}$). When an axial current ($I$) from a pulsed power generator is applied, the gas is ionized and radially accelerated inward by the current self-generated azimuthal magnetic field ($B_\theta$), thus compressing the preembedded $B_z$.

If the plasma acts as a perfect conductor, i.e., the axial magnetic field is frozen into the plasma, then $B_z$ increases as $B_{z0}(R(t=0)/R(t))^2$, where $R(t)$ is the plasma radius at time $t$. However, there are arguments that this simple picture of flux compression does not entirely hold. For example, inside the metal electrodes, located at each end of the pinch, the flux is truly frozen at its initial value. Hence, the magnetic field lines exit the electrodes at their initial radial position and are pulled into the plasma as it implodes. In other words, the plasma must have a finite resistivity which allows the field to diffuse near the two electrodes. Therefore one of the main goals of the present work was to measure the axial magnetic field distribution and evolution.

The efficiency of the compression is also affected by instabilities. It is known that the axial magnetic field may stabilize the magneto-Rayleigh-Taylor instability (MRTI) [12, 20]. Therefore, one of the objectives of the present work is the investigation of the effects of the axial magnetic field on plasma instabilities.

Table 1.1 summarizes the parameters of the previous and current experiments of $B_z$-flux compression by an imploding plasma. From the table we see that the subject of $B_z$ compression by plasma implosion was studied in a wide range of experimental parameters. Most of the laboratories employed a gas-puff load, as in our experiment, since it facilitates the systematic study of the compression due to the possibility of high repetition rate of the plasma generation. Here we choose a relatively low value for the current (and thus low value for $B_{z0}$) that allows for the application of UV-Vis spectroscopy. It is assumed that the fundamental physical processes of the

**Table 1.1** Experimental parameters of previous and current studies of the $B_z$ compression by Z-pinch implosion. The studies are presented in a chronological order

| Institution, driver | Load configuration | Load composition | $I_0$ MA | $t_{rise}$ ns | $R_0$ mm | $B_{z0}$ T | References |
|---|---|---|---|---|---|---|---|
| UC Irvine | Single-shell gas-puff | He, Ar, Kr, Xe | 0.5 | 1250 | 20 | $0 \leq B \leq 1.5$ | [12] |
| Sandia, Proto-II | Single-shell gas-puff | Ne | 7.5 | 60 | 12.5 | $5.5 \leq B \leq 10$ | [13] |
| Imperial college | Quartz tube, Pyrex tube with gas | $H_2$, $D_2$, He, Ne, Ar, Kr | 0.53 | 2000 | 15, 26.7 | $0 \leq B \leq 1$ | [14] |
| HCEI, IMRI-5 | Double-shell gas-puff | Ne | 0.4 | 430 | 22, 30, | $0 \leq B \leq 0.2$ | [21] |
| Troitsk, Angara-5 | Wire array | Tungsten | 2 | 800 | 10 | $0 \leq B \leq 1.4$ | [15] |
| HCEI GIT-12 | Double-shell gas-puff | Ar | 2.5 | 300 | 40 | $0 \leq B \leq 0.6$ | [22] |
| HCEI, MIG | Single-shell gas-puff | $D_2$, $N_2$, Ne, Ar | 1 | 150 | 14 | 2.5 | [17] |
| Sandia, Z | Solid liner | Beryllium | 19 | 100 | 2.5 | 10 | [23] |
| WIS | Single-shell gas-puff | Ar | 0.3 | 1600 | 19 | $0 \leq B \leq 0.4$ | [24, 25] |
| HCEI, IMRI-5 | Metallic gas-puff | Bismuth | 0.45 | 450 | $\sim 20$–30 | $0.15 \leq B \leq 1.35$ | [26, 27] |
| Michigan university, MAIZE | Metallic foil | Aluminum | 0.58 | 250 | 3 | $0 \leq B \leq 2$ | [28, 29] |
| Cornell, university, COBRA | Double-shell on jet gas-puff | Ar | 1 | 200 | 30 | $0 \leq B \leq 0.6$ | [30, 31] |
| University of Nevada, Reno ZEBRA | Single-shell on jet gas-puff | Ar, Kr $D_2$ | 1 | 100 | 11 | $0 \leq B \leq 0.15$ | [32–34] |

magnetic field and magnetized plasma compression are the same over a wide range of experimental parameters.

## 1.2  Z-Pinch Principle and Practical Considerations

In the following, a short introduction into Z-pinch physics is given. For a deeper introduction, there are several reviews and a textbook available [35–40]. For the theoretical background we mainly follow the recent review of gas-puff Z-pinches by Giuliani and Commisso [35].

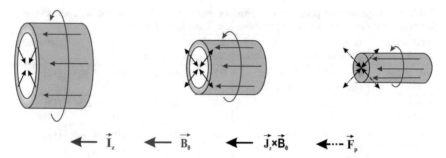

**Fig. 1.1** Main stages of Z-pinch evolution without axial magnetic field. Left, middle, and right figures are respectively, the breakdown, implosion, and stagnation stages. $\vec{I}_z$ is the axial current, $\vec{B}_\theta$ is the azimuthal magnetic field generated by the axial current, $\vec{J}_z \times \vec{B}_\theta$ is the inward Lorentz force, and $\vec{F}_p$ is the outward force due to the thermal pressure gradient

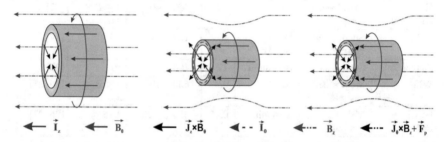

**Fig. 1.2** Main stages of Z-pinch evolution with pre-embedded axial magnetic field. Left, middle, and right figures are respectively, the breakdown, implosion, and stagnation stages

The Z-pinch is a cylindrical plasma configuration through which an axial electric current is driven. The current is usually generated by a pulsed-power generator. The interaction of the axial current with the self-generated azimuthal magnetic field (Lorentz force $\vec{j} \times \vec{B}$) results in the plasma implosion toward the symmetry axis (typical implosion times are in the range 100 ns–1 µs). During the implosion, the plasma acquires kinetic and thermal energy due to the Lorentz force and Ohmic heating and eventually looses most of its energy to electromagnetic radiation. Different stages of the Z-pinch implosion without (classical Z-pinch) and with $B_z$ are depicted in Figs. 1.1 and 1.2.

The usual scenario of a Z-pinch evolution consists of three main stages:

- First stage—generation of plasma: breakdown, avalanche ionization and initial heating of a gas. In some cases, prior to the breakdown occurs also vaporization of solids (metallic wires, foil, fibers).
- Second stage—implosion: further ionization and heating of the plasma by resistive heating ($\eta \cdot j^2$, $\eta$ is the plasma resistivity, and $j$ is the current density), and by adiabatic compression due to the Lorentz force ($j_z \cdot B_\theta$). Inward acceleration of the plasma shell due to the Lorentz force. In the case of implosion with $B_z$, there

is also compression of $B_z$ that results in increase of the magnetic energy stored in the $B_z$-field.

- Third stage—stagnation: the kinetic energy of the imploding plasma is converted into thermal energy and subsequently lost through radiation. In the case of implosion with $B_z$, part of the kinetic energy is transformed into the $B_z$-field energy. It is important to note, Kroupp et al. [41] showed that in classical Z pinches at stagnation the kinetic energy of the implosion is first converted into the hydrodynamic motion, probably turbulent, and later is dissipated into the thermal energy due to the ion viscosity.

Z-pinches are studied in a wide range of initial conditions and plasma parameters. The range of typical values for these parameters is listed below:

- Stored energy in capacitors of a pulsed-power generator: few tens J—22 MJ (Sandia Z machine [42]).
- Peak current of a pulsed-power generator: few tens of kA—26 MA (Sandia Z machine).
- Rise-time of the current pulse (correlated with the plasma implosion time): few tens ns—few $\mu$s.
- Initial radius of the plasma: $\sim$1–50 mm (note that Z-pinches with a single wire load, the wire radius is $\ll$1 mm, however, during the vaporization stage it expands to $\sim$ mm scale).
- Length of plasma (anode-cathode gap): few mm–20 mm.
- Initial axial magnetic field: 0.1–10 T.
- Electron density, $n_e$: $>$ $10^{17}$–few $10^{23}$ cm$^{-3}$.
- Electron temperature, $T_e$: few eV–10 keV.

There are a variety of load configurations from which Z-pinch plasma is generated, the most common configurations studied today are: (i) gas-puffs [35]; (ii) wire-arrays [36], and (iii) solid liners [43]. In the present experiment, a gas-puff load is used due to several advantages that are important for systematic Z-pinch investigation:

(i) possibility of multiple discharges without breaking vacuum and relatively high repetition rate for the shots ($\equiv$discharges). For example, in the present experiment, the vacuum chamber is opened for cleaning every $\sim$100 shots and it is possible to perform $\sim$4 discharges per hour;

(ii) High flexibility in controlling the initial distribution of the load line-mass (mass per unit length). For example, in the present experiment the line-mass can be easily varied from $<$ 1 $\mu$g/cm up to $\sim$100 $\mu$g/cm by changing the initial gas density or gas species. Additionally, gas-puff systems with several nozzles allow for the control of the load radial density distribution that is important for mitigation of the magnetic Rayleigh–Taylor instability [44, 45];

(iii) Ease to use hydrogen isotopes as a Z-pinch load for fusion research purposes (or neutron sources), since they are in the gas phase in a wide range of pressures and temperatures.

Besides the advantages described above, there are also several challenges and limits when utilizing of gas-puff loads. Below, these challenges are discussed together with their solutions in the present experiment.

(i) Axial and azimuthal uniformity of the initial gas distribution. To minimize the radial divergence of the gas along the A-K gap, the nozzle of the gas-puff system is designed to generate uniform, collimated supersonic flow using converging-diverging geometry as shown in Fig. 2.2 (such geometry is called de-Laval nozzle). The uniformity and collimation can be disrupted in the boundary layers at the inner and outer radii of the nozzle due to the friction of the gas-flow with nozzle walls (for more details see [46]).

(ii) Backscattering of the expanding gas from an opposite electrode into the A-K gap. To minimize the amount of backscattered gas, the opposite electrode is made of wire-mesh (or some other semi-transparent structure) and the discharge is initiated before a significant amount of the backscattered gas fills the A-K gap.

(iii) Uniformity of the gas-breakdown. To achieve uniform breakdown, the gas inside the A-K gap is usually pre-ionized by UV radiation or an electron beam. In addition, the opposite to the nozzle electrode has a knife-edge attached to the wire-mesh (see Fig. 2.1).

(iv) Gas-puff loads have some limitation on the maximum particle density they can generate. The particle density is set by the initial pressure supplied to the gas-puff system and the gas expansion in the nozzle. Most of the gas-puff systems operate in the $10^{16} - 10^{18}$ cm$^{-3}$ range, but some special designs can reach particle densities of up to $\sim 10^{20}$ cm$^{-3}$ in the A-K gap. In general, high densities might be required in very-high-current devices together with the requirement of small initial radius.

### 1.2.1   Magneto-Hydrodynamic Description of Z-Pinch Dynamics

The Z-pinch dynamics is usually described using magneto-hydrodynamic (MHD) models. Magneto-hydrodynamics studies the dynamics of conducting gases and fluids in electromagnetic fields where the fields and fluids equations are coupled. The presence of an electromagnetic field where a conducting fluid is in motion results in an induction of electrical currents inside the medium. The induced currents, on the one hand, interact with the magnetic field in the form of the Lorentz force that changes the hydrodynamic motion of the fluid. On the other hand, the induced electrical currents change the electromagnetic field. MHD attempts to describe this coupled evolution of the conducting fluid (plasma) and the electromagnetic field, namely, beside the plasma equation of motion, MHD includes also Maxwell's equations and Ohm's law.

Here, we present the equations used for the single fluid MHD model without including the radiation and ionization processes. The radiation losses and internal energy due to ionization and excitations, which are usually important in Z-pinch models, are considered here using lumped terms in the energy balance equation. We follow mainly the MHD description given in Giuliani and Commisso [35], and Braginskii [47]. Note,

(i) in both references the equations are given in cgs units, while we present them using the SI units due to the convenience of using Ampere and Volt units in calculations that appear later in the text.
(ii) Terms in equations that include the Boltzmann constant ($k_B$) explicitly, have a temperature in Kelvin (K). When temperatures are expressed in electron-volts (eV), it is explicitly written next to the variable. Similarly, densities are expressed in the $m^{-3}$ unit except when the $cm^{-3}$ is explicitly written next to the variable. Here, the transport coefficients equations are given also using eV and $cm^{-3}$, since these units are convenient and are widely used by the plasma community.
(iii) The equations are given for fully ionized, not magnetized plasma, i.e., $\omega_{ce} < \nu_{ei}$ ($\omega_{ce}$—the electron cyclotron angular frequency, $\nu_{ei}$—electron-ion collision frequency). In general, Z pinches might have regions where the electrons are magnetized, for such case the transport (electrical current, momentum, heat) in the plasma has to be distinguished between transport along the B-field direction and perpendicular to it (for more details see [47]).

The one fluid MHD equations are:

• Absence of magnetic monopoles:

$$\nabla \cdot \vec{B} = 0 \tag{1.1}$$

• Faraday's law:

$$\frac{\partial \vec{B}}{\partial t} = -\nabla \times \vec{E} \tag{1.2}$$

• Ampere's law without the electric displacement field term. This term can be neglected when the plasma velocity $v \approx v_i$ ($v_i$—ion velocity) $\ll c$ (speed of light). This condition is satisfied by all Z-pinches.

$$\nabla \times \vec{B} = \mu_0 \vec{j} \tag{1.3}$$

$\vec{j}$ is the current density, $\mu_0$ is the vacuum permeability.
• Generalized Ohm's law (commonly used to express electric field in plasma):

$$\vec{E} = -\vec{v} \times \vec{B} + \eta \vec{j} + \frac{\vec{j} \times \vec{B}}{en_e} - \frac{\nabla p_e}{en_e} + \frac{\vec{j} \cdot \vec{R}_T}{en_e} + \frac{m_e}{e^2 n_e} \frac{d\vec{j}}{dt} \tag{1.4}$$

with

$$\eta = \alpha_e \frac{m_e \nu_{ei}}{n_e e^2} = 3 \times 10^{-12} \cdot \alpha_e \frac{m_e \bar{Z} ln(\Lambda)}{e^2 T_e^{3/2}(\text{eV})} = 10^{-4} \cdot \alpha_e \frac{\bar{Z} ln(\Lambda)}{T_e^{3/2}(\text{eV})} (\Omega \cdot \text{m}) \quad (1.5)$$

$$\nu_{ei} = 3 \times 10^{-6} \cdot \frac{n_e(\text{cm}^{-3}) \bar{Z} ln(\Lambda)}{T_e^{3/2}(\text{eV})} \quad (1.6)$$

$$\alpha_e \approx \frac{1 + 1.2\bar{Z} + 0.22\bar{Z}^2}{1 + 2.97\bar{Z} + 0.75\bar{Z}^2} \quad (1.7)$$

$$\vec{R}_T = -\alpha_q n_e k_B \nabla T_e \quad (1.8)$$

$\eta$ is the plasma resistivity, $ln(\Lambda)$ is the Coulomb logarithm, $\bar{Z}$ is the average charge state, $\nu_{ei}$ is electron-ion collision frequency, $\alpha_e$ is a factor introduced due to the electron-electron collision effects that tend to relax the electron distribution function into the velocity shifted Maxwellian distribution. Equation 1.7 is an approximation of the $\alpha_e$ factor that is 0.5129 for $\bar{Z} = 1$ and $3\pi/32$ for $\bar{Z} \to \infty$ (for more details see [48]). $R_T$ is a thermal force that originates from the $v^{-3}$ dependence of the Coulomb collision cross section on the electron velocity, resulting in a friction force between ions and electrons when $\nabla T_e \neq 0$. $\alpha_q$ is a factor that depends on $\bar{Z}$ ($\alpha_q$ for selected values of $\bar{Z}$ is given in Table 1.2). To convert the generalized Ohm's law from SI to cgs units one has to replace the $\vec{v} \times \vec{B}$ and $\vec{j} \times \vec{B}/(en_e)$ terms in Eq. 1.4 by $\vec{v} \times \vec{B}/c$ and $\vec{j} \times \vec{B}/(cen_e)$, respectively, and use $\eta(\text{sec}) = 1.1 \times 10^{-10} \cdot \eta$ ($\Omega$·m). To convert equations with $T_e$ expressed in eV into equations with $T_e$ expressed in Kelvin, substitute $T_e(\text{eV})$ by $k_b(\text{eV/K})T(\text{K}) \approx 8.62 \times 10^{-5} \cdot T(\text{K})$.

- Continuity equation:

$$\frac{\partial \rho}{\partial t} + \nabla \cdot \rho \vec{v} = 0 \quad (1.9)$$

$\rho \approx n_e M_i / \bar{Z}$ is the plasma mass density, $M_i$—ion mass.

- Equation of motion (repeated subscripts indicate a sum over all the three coordinates):

$$\rho \frac{\partial \vec{v}}{\partial t} + \rho(\vec{v} \cdot \nabla)\vec{v} = -\nabla(p_i + p_e) - \frac{\partial \Pi_{\alpha\beta}}{\partial x_\beta} + \vec{j} \times \vec{B} \quad (1.10)$$

with

$$\Pi_{\alpha\beta} = -\eta_{visc} \left( \frac{\partial v_\alpha}{\partial x_\beta} + \frac{\partial v_\beta}{\partial x_\alpha} - \frac{2}{3} \delta_{\alpha\beta} \nabla \cdot \vec{v} \right) \quad (1.11)$$

$$\eta_{visc}(\text{Pa} \cdot \text{s}) \approx 0.96 \cdot n_i k_B T_i \tau_{ii} \approx 3.2 \times 10^{-6} \frac{A^{1/2} T_i^{5/2}(\text{eV})}{Z^4 ln(\Lambda)} \quad (1.12)$$

$$\tau_{ii} \approx 2 \times 10^7 \frac{A^{1/2} T_i^{3/2}(\text{eV})}{Z^4 n_i(\text{cm}^{-3}) ln(\Lambda)} \quad (1.13)$$

$n_i (= n_e/\bar{Z})$, $T_i$, and $p_i = n_i k_B T_i$ are the ion density, temperature and pressure, respectively (note, $T_i$ is in Kelvin and not in electronvolt unless specified otherwise). $\Pi_{\alpha\beta}$ is the ion viscosity stress tensor, $\eta_{visc}$ is the ion viscosity coefficient, $A$ is the ion mass expressed in proton mass, and $\tau_{ii}$ is the ion-ion collision time. To obtain the equation of motion in cgs units, one replaces $\vec{j} \times \vec{B}$ by $\vec{j} \times \vec{B}/c$, and use $\eta_{visc}(\text{poise}) = 10 \times \eta_{visc}(\text{Pa} \cdot \text{s})$.

- Ion thermal energy equation:

$$\frac{\partial}{\partial t}\left(\frac{3}{2}n_i k_B T_i\right) + \nabla \cdot \left(\frac{3}{2}n_i k_B T_i \vec{v}\right) = -p_i \nabla \cdot \vec{v} - \Pi_{\alpha\beta}\frac{\partial v_\alpha}{\partial x_\beta} - \nabla \cdot (\kappa_i \nabla T_i) + Q_{ie}$$

(1.14)

where

$$\kappa_i = 3.9 \cdot \frac{n_i k_B^2 T_i(\text{K})\tau_{ii}}{M_i} \approx 0.1 \cdot \frac{T_i^{5/2}(\text{eV})}{\bar{Z}^4 A^{1/2} ln(\Lambda)}(\text{J} \cdot \text{s}^{-1} \cdot \text{K}^{-1} \cdot \text{m}^{-1})$$

(1.15)

$$Q_{ie} = \frac{3m_e}{M_i}n_e \nu_{ei} k_B (T_e - T_i) =$$

(1.16)

$$= 7.9 \times 10^{-22} \cdot \frac{n_e^2(\text{cm}^{-3})\bar{Z}ln(\lambda)}{T_e^{3/2}(\text{eV})}(T_e(\text{eV}) - T_i(\text{eV}))(\text{J} \cdot \text{m}^{-3} \cdot \text{s}^{-1})$$

$\kappa_i$ is the ions thermal conductivity coefficient ($\kappa_i(\text{erg} \cdot \text{sec}^{-1} \cdot \text{K}^{-1} \cdot \text{cm}^{-1}) = 10^5 \times \kappa_i(\text{J} \cdot \text{s}^{-1} \cdot \text{K}^{-1} \cdot \text{m}^{-1})$, $Q_{ie}$ is the heat acquired by the ions due to the collisions with electrons ($Q_{ie}(\text{erg} \cdot \text{cm}^{-3} \cdot \text{sec}^{-1}) = 10 \times Q_{ie}(\text{J} \cdot \text{m}^{-3} \cdot \text{s}^{-1})$). The 1st term at the r.h.s of Eq. 1.14 represents heating by adiabatic compression; the 2nd term represents viscous heating; the 3rd term represents the ion heat conduction; and the 4th term represents the heat transfer from electrons to ions.

- Electron thermal energy equation:

$$\frac{\partial}{\partial t}\left(\frac{3}{2}n_e k_B T_i + n_i \epsilon_x\right) + \nabla \cdot \left(\frac{3}{2}n_e k_B T_e \vec{v}_e\right) + \nabla \cdot (n_i \epsilon_x \vec{v}) =$$

$$= -p_e \nabla \cdot \vec{v}_e - \nabla \cdot \left(-\kappa_e \nabla T_e - \alpha_q k_B T_e \vec{j}/e\right) - Q_{ie} + \eta j^2 + \frac{\vec{j} \cdot \vec{R}_T}{en_e} - W_{rad}$$

(1.17)

with

$$\kappa_e = \alpha_\kappa \frac{n_e k_B^2 T_e}{m_e \nu_{ei}} = \alpha_\kappa 0.85 \frac{T_e^{5/2}(\text{eV})}{\bar{Z}ln(\Lambda)}(\text{J} \cdot \text{s}^{-1} \cdot \text{K}^{-1} \cdot \text{m}^{-1})$$

(1.18)

$\epsilon_x$ is the sum of the ionization and excitation energy per ion. This term is included in the electron thermal energy equation since electrons loose (gain) their energy through ionization (recombination) and excitation (de-excitation) of the ions. $\vec{v}_e \approx \vec{v} - \vec{j}/(en_e)$ is the electrons velocity, and $\vec{v}_{rel} = -\vec{j}/(en_e)$ is the relative velocity

**Table 1.2** $\alpha_\kappa$ and $\alpha_q$ factors for different ionization stages

| $\bar{Z}$ | 1 | 2 | 3 | 4 | $\infty$ |
|---|---|---|---|---|---|
| $\alpha_\kappa$ | 3.16 | 4.9 | 6.1 | 6.9 | 12.5 |
| $\alpha_q$ | 0.71 | 0.9 | 1 | 1.1 | 1.5 |

of the electrons and ions. $\kappa_e$ is the electrons thermal conductivity coefficient, $\alpha_\kappa$ is a factor that depends on $\bar{Z}$ and is given in Table 1.2 for selected values of $\bar{Z}$. The term $\alpha_q k_B T_e \vec{j} / e$ (sometimes appears as $\alpha_q p_e \vec{j} / (e n_e)$ or $\alpha_q n_e k_B T_e \vec{v}_{rel}$) represents the heat flux due to the electrical current. It originates from the $v^{-3}$ dependence of the Coulomb collision cross section on the electron velocity, therefore, the current is mostly carried by the faster electrons. This leads to the heat flow in the rest frame of the electrons in the direction opposite to the current (for more details see [47]). The term $\eta j^2$ represents resistive (Joule) heating by the electrical current, $\vec{j} \cdot \vec{R}_T / (e n_e)$ represents heating due to the thermal force, and $W_{rad}$ represents energy loss through radiation.

For the detailed derivations of the equations presented above and an extension of these equations to the case when the plasma is magnetized, see [47]. For different forms of the energy and momentum equations (1.10), (1.14), (1.17) that are useful for understanding the energy partition (electromagnetic, thermal, kinetic) and energy evolution in a Z-pinch see [35].

### 1.2.2 Ideal MHD Model

Applying and solving Eqs. 1.1–1.17 for the Z-pinch evolution even by using computer simulations, is extremely difficult, since there are many physical processes that has to be considered. These include resistive and adiabatic heating versus radiation and heat conduction, magnetic field diffusion versus convection with plasma flow, development of instabilities and turbulences, ionization dynamics, and more. Many of these processes are coupled and therefore the solution has to be self-consistent. Due to the complexity of the physics involved in the plasma implosion, reliable, detailed experimental data is essential for the testing of different MHD codes.

Due to the high complexity of the physics involved, it is sometimes useful to employ a simplified model that gives some insights into the Z-pinch dynamics. Such simplified model makes use of the ideal MHD equations where collisional effects between the plasma particles are neglected, consequently, the conductivity is infinite, no radiation, no viscosity, and no interaction with the electrodes are considered. The set of the ideal MHD equations is:

$$\frac{\partial \rho}{\partial t} + \nabla \cdot \rho \vec{v} = 0, \qquad \text{mass continuity} \qquad (1.19)$$

$$\rho \frac{d\vec{v}}{dt} = -\nabla p + \vec{j} \times \vec{B}, \qquad \text{equation of motion} \qquad (1.20)$$

$$\frac{\partial \vec{B}}{\partial t} = -\nabla \times \vec{E}, \qquad \text{Faraday's Law} \qquad (1.21)$$

$$\nabla \times \vec{B} = \mu_0 \vec{j}, \qquad \text{Ampere's law} \qquad (1.22)$$

$$\vec{E} + \vec{v} \times \vec{B} = 0, \qquad \text{Ohm's law} \qquad (1.23)$$

$$\frac{d}{dt}\left(\frac{p}{\rho^{\gamma}}\right) = 0, \qquad \text{equation of state} \qquad (1.24)$$

where $p = p_e + p_i$—plasma pressure, $\gamma$—adiabatic constant (in plasmas usually $\gamma = 5/3$).

This simplified set of equations allows for the study of Z-pinch dynamics by one-dimensional (only radial dependence), self-similar solutions where the variables are assumed to be separable functions of time and radius. Although these analytical solutions don't include some important physical processes, like the development of instabilities, $B$-field diffusion, radiation and more, they still provide valuable qualitative insight and guidance into the general Z-pinch dynamics and the relations between the magnetic field, plasma, and flow parameters. Detailed discussion on the applicability of ideal MHD equations to Z-pinch conditions, and on the different types of self-similar solutions for ideal Z-pinch dynamics can be found in M. A. Liberman's et al., book 'Physics of high-density Z-pinch plasmas' [37].

Another important analytical solution that can be derived using ideal MHD equations describes the Z-pinch plasma in force equilibrium and $\vec{v}(r, t) = 0$. This solution considers the balance between the Lorentz forces due to the axial and azimuthal current distribution $j_z(r)$ and $j_\theta(r)$, respectively, and the thermal pressure gradient in an infinite plasma cylinder. Using the equation of motion (Eq. 1.20) and Ampere's law (Eq. 1.22), the equilibrium can be described by:

$$\frac{dp}{dr} = -\frac{1}{2\mu_0}\frac{d}{dr}(B_z^2 - B_{z0}^2) - \frac{1}{2\mu_0 r^2}\frac{d}{dr}(r^2 B^2) \qquad (1.25)$$

This leads to the relation between the average pressure $\langle p \rangle$ inside the plasma column, the total current $I$, the initial axial magnetic field $B_{z0}$, the initial plasma radius $R_0$ (before the start of the compression), and the equilibrium plasma radius $R$:

$$\langle p \rangle \equiv \frac{2\pi}{\pi R^2}\int_0^R rp(r)dr = \frac{\mu_0 I^2}{8\pi^2 R^2} - \frac{1}{2\mu_0}B_{z0}^2\left(\frac{\langle B_z^2 \rangle}{B_{z0}^2} - 1\right) = \frac{\mu_0 I^2}{8\pi^2 R^2} - \frac{1}{2\mu_0}B_{z0}^2\left(\alpha^2\frac{R_0^4}{R^4} - 1\right) \qquad (1.26)$$

i.e. the average plasma pressure is equal to the azimuthal magnetic pressure at the outer radius of the plasma ($B^2(r = R)/(2\mu_0)$) minus the average axial magnetic pressure ($\langle B_z^2 \rangle$ is defined similarly to $\langle p \rangle$). $\alpha$ is the parameter varying in the range $R^2/R_0^2 < \alpha < 1$ that characterizes the confinement of the axial magnetic flux.

The first theoretical study of a pinch in force equilibrium (without $B_z$) appeared in Bennett's paper [49] from 1934, where he obtained a relation (today is called "Bennett relation") between the total axial current ($I$), the number of ions per unit length along the axis ($N$), and $T_B$ ("Bennett temperature") that is the mean temperature weighted by density:

$$(1 + \bar{Z})Nk_bT_B = \frac{\mu_0 I^2}{8\pi} \tag{1.27}$$

$$k_bT_B = \frac{2\pi}{N(1 + \bar{Z})} \int_0^R rp(r)dr \tag{1.28}$$

where $p(r)$ is the plasma pressure (electron+ion), $R$ is the outer radius of the plasma column, and $\bar{Z}$ is the average charge state. Assuming an isothermal pinch with equal electron and ion temperatures, a more convenient form of the Bennett relation can be used [37]:

$$T(\text{keV}) = 3.12(I/\text{MA})^2(1 + \bar{Z})^{-1}(N/10^{18}\text{cm}^{-1})^{-1} \tag{1.29}$$

Note, Eqs. 1.26 and 1.27 are equivalent forms to describe the relation between plasma and magnetic field parameters.

### 1.2.3  Snow-Plow Model

To estimate the evolution of the plasma implosion velocity, and the time and radius of the stagnation as a function of $B_{z0}$, we use an even more simplified model called "Snow-plow model" [35]. This is a 0D model, in which the current-carrying imploding plasma is assumed to be an infinitely thin cylindrical shell. As the plasma shell implodes, it sweeps up material to the velocity of the shell. This model doesn't include the thermal pressure term in the momentum equation, since the outward force due to thermal pressure in our strongly radiating plasma is much smaller than the Lorentz forces during the implosion stage. Here, we use the "Snow-plow" model to describe the implosion of a plasma shell that in addition to sweeping material as it implodes, it also compresses an axial magnetic field. The current evolution is then described by the parameters of the pulsed-power generator coupled to the plasma inductance evolution that is calculated using plasma dimensions evolution. Since the plasma-shell has infinite conductivity, that implies conservation of the axial magnetic flux within its radius. This leads to the following dependence of $B_z(t)$ on the plasma radius $R(t)$: $B_z(t) = B_{z0}\left(R_0^2/R(t)^2\right)$, where $R_0$ is the outer radius of the gas shell (and the plasma radius at $t = 0$) and $R(t)$ is the plasma radius at time $t$.

The equation of motion for the plasma shell is:

for $R(t) > R_{in}$

$$\frac{d}{dt}\left(m(t)\frac{dR}{dt}\right) = \frac{m_0}{\pi(R_0^2 - R_{in}^2)}\frac{d}{dt}\left(\pi(R_0^2 - R(t)^2)\frac{dR}{dt}\right) =$$
$$= -\frac{\mu_0 I(t)^2}{4\pi R(t)} - \frac{B_{z0}^2}{\mu_0}\pi R(t)\left(1 - \frac{R_0^4}{R(t)^4}\right), \tag{1.30}$$

for $R(t) < R_{in}$

$$m_0\frac{d^2 R}{dt^2} = -\frac{\mu_0 I(t)^2}{4\pi R(t)} - \frac{B_{z0}^2}{\mu_0}\pi R(t)\left(1 - \frac{R_0^4}{R(t)^4}\right), \tag{1.31}$$

where $m_0$(kg/m) is the total line-mass (gas-shell mass divided by the shell length), $m(t)$ is the imploding line-mass at time $t$, $R_{in}$ is the inner radius of the gas shell, and $I(t)$ is the plasma current at time $t$. The first term on the right hand side of the equations represents the inward force exerted on the plasma shell by the $B_\theta$ pressure. The second term represents the outward force exerted by the $B_z$ pressure. The term $m_0/(\pi(R_0^2 - R_{in}^2))$ represents the initial mass density per length. It is assumed to be uniform and non-zero in the range $R_{in} < R < R_0$ and zero otherwise. Note, in order to include also thermal pressure force in the equation of motion, the terms $k_B T(Z+1)m(t)/M_{ion}/R(t)$ and $k_B T(Z+1)m_0/M_{ion}/R(t)$ has to be added in the r.h.s. of Eqs. 1.30 and 1.31, respectively ($M_{ion}$ is the ion mass).

The current evolution is calculated by:

$$\frac{d^2}{dt^2}(L_{total} \cdot I) + R_{PPS} \cdot \frac{dI}{dt} - \frac{I}{C} = 0, \tag{1.32}$$

where $C$ and $R_{PPS}$ are, respectively, the capacitance and resistance of the pulse-power system (according to the model, the plasma resistance is zero). $L_{total} = L_{PPS} + L_{plasma}$, where $L_{PPS}$ is the inductance of the pulse-power machine without plasma and $L_{plasma}$ is the inductance of the plasma column given by:

$$L_{plasma} = \frac{\mu_0 l}{2\pi}\ln\left(\frac{R_{return}}{R(t)}\right) \tag{1.33}$$

where $l$ is the length of the plasma column and $R_{return}$ is the return current radius.

To demonstrate the effects of an axial magnetic field on the plasma implosion with typical discharge parameters and initial conditions, several calculations of the $R(t)$, $B_z(t)$ and $B_\theta(t)$ are presented in Figs. 1.3, 1.4, 1.5, 1.6 and 1.7. The discharge parameters and initial conditions used in the calculations are given in Table 1.3. Stagnation is defined as the time when the plasma radius reaches its minimum. At this time we expect the maximum compression of $B_z$. Table 1.4 summarizes the calculated values of the stagnation time and the corresponding plasma radius, the maximum $B_z$

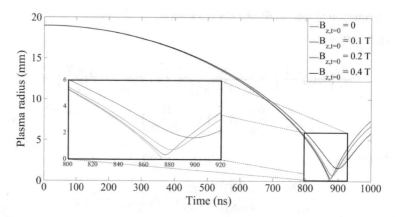

**Fig. 1.3** Calculated plasma radius evolution $R(t)$ for $m_{0,Ar} = 30\,\mu$g/cm for four different $B_{z0}$. The insert highlights the plasma radius evolution for different $B_{z0}$ at times close to the stagnation

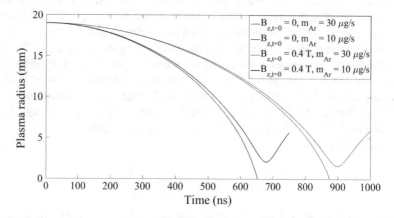

**Fig. 1.4** Plasma radius evolution $R(t)$ calculated for two different mass-loads per length $m_{0,Ar} = 30\,\mu$g/cm and $m_{0,Ar} = 10\,\mu$g/cm, and for $B_{z0} = 0$ and $B_{z0} = 0.4$ T

(i.e $B_z$ at stagnation), and $B_\theta$ at stagnation for different initial conditions. Figure 1.3 presents the results of the plasma radius calculations for $m_{0,Ar} = 30\,\mu$g/cm and four different $B_{z0}$. It is seen that in the range of $B_{z0} \leq 0.4$ T, used in the present experiment, the effect of $B_z$ on the plasma radius evolution during compression is expected to be significant only at the late stages of the compression (see insert in Fig. 1.3), whereas the times of stagnation differ by $<3\%$ between the lowest and highest $B_{z0}$. We note that the calculation for implosion without $B_z$ proceeds until $R = 0$ and is terminated from this point, whereas for $B_{z0} > 0$ the plasma radius increases after the stagnation due to the plasma inertia effect, explained below.

Figure 1.4 compares the plasma radius evolution between the implosions with two different initial gas-shell masses per length $m_{0,Ar} = 30\,\mu$g/cm and $m_{0,Ar} = 10\,\mu$g/cm, for $B_z = 0$ and $B_z = 0.4$ T. We see that $3\times$ increase of $m_{0,Ar}$ delays

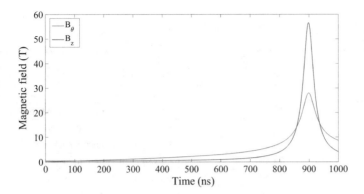

**Fig. 1.5** Calculated $B_\theta$ and $B_z$ evolution for $m_{0,Ar} = 30\ \mu g/cm$ and $B_{z0} = 0.4\ T$

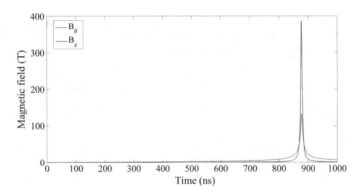

**Fig. 1.6** Calculated $B_\theta$ and $B_z$ evolution for $m_{0,Ar} = 30\ \mu g/cm$ and $B_{z0} = 0.1\ T$

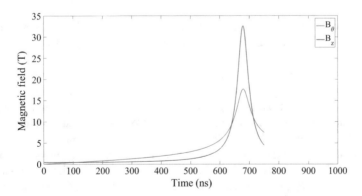

**Fig. 1.7** Calculated $B_\theta$ and $B_z$ evolution for $m_{0,Ar} = 10\ \mu g/cm$ and $B_{z0} = 0.4\ T$

**Table 1.3** Initial conditions and discharge parameters

| Parameter | Value | Description |
|---|---|---|
| $B_{z0}$ | 0, 0.1, 0.2, 0.4 T | Initial axial magnetic field |
| $R_0$ | 19 mm | Outer radius of the initial gas distribution |
| $R_{in}$ | 7 mm | Inner radius of the initial gas distribution |
| $m_0$ | 10, 30 μg/cm | Gas-shell mass divided by the shell length |
| $V_0$ | 23 kV | Initial voltage of the capacitors |
| $C$ | 16 μF | Pulse-power system capacitance |
| $L_{PPS}$ | 55 nH | Pulse-power system inductance |
| $R_{PPS}$ | 15 mΩ | Pulse-power system resistance |
| $R_{return}$ | 125 mm | Return current radius |

**Table 1.4** Summary of the theoretical calculations of $B_z$ compression for different initial conditions: (1) mass per length of argon gas-shell, (2) initial axial $B$-field, (3) time of stagnation, (4) radius of stagnation, (5) $B_z$ at stagnation, (6) compression factor of $B_z$, (7) $B_\theta$ at stagnation

| $m_{0,Ar}$ μg/cm | $B_{z0}$ T | $t_{stagnation}$ ns | $R_{stagnation}$ mm | $B_{z,stagnation}$ T | $\frac{B_{z,stagnation}}{B_{z0}}$ | $B_{\theta,stagnation}$ T |
|---|---|---|---|---|---|---|
| 10 | 0 | 651 | | | | |
| 10 | 0.4 | 679 | 2.1 | 32.5 | 81 | 17.5 |
| 30 | 0 | 874 | | | | |
| 30 | 0.1 | 877 | 0.3 | 388 | 3880 | 132 |
| 30 | 0.2 | 882 | 0.7 | 153 | 765 | 61.5 |
| 30 | 0.4 | 899 | 1.6 | 57 | 142 | 28 |

the stagnation time by ~30% and decreases the stagnation radius (relevant only for implosions with $B_z$) by ~25%. Figures 1.5 and 1.6 present $B_\theta$ and $B_z$ evolution for $B_{z0} = 0.1$ T and $B_{z0} = 0.4$ T. The assumed argon mass-load per length is $m_{0,Ar} = 30$ μg/cm. In both figures, $B_z$ becomes larger than the compressing $B_\theta$ at times close to stagnation. This phenomenon is expected since the plasma shell has non zero velocity, when the magnetic pressure due to $B_z$ becomes equal to the magnetic pressure due to $B_\theta$. Therefore, its inertia continues to compress the $B_z$-field, resulting in $B_z > B_\theta$. Another effect seen in these two figures is that for lower $B_{z0}$, $B_{z,stagnation}$ is larger. This can be explained by two factors. The first is the significant increase of $B_\theta$ pressure close to stagnation for lower $B_{z0}$ since the plasma radius reaches smaller values before the $B_z$ pressure stops the acceleration. The second factor is the larger inertia gained by the plasma-shell since the $B_z$ pressure acting outward is smaller for lower $B_{z0}$ during most of the implosion stage.

Figure 1.7 presents the evolution of $B_z$ and $B_\theta$ for $B_{z0} = 0.4$ T and $m_{0,Ar} = 10$ μg/cm. Comparing the $B_{z,stagnation}$ in Figs. 1.7 and 1.5, which have the same $B_{z0}$, but different $m_{0,Ar}$, we see that $B_{z,stagnation}$ is larger for higher $m_{0,Ar}$. This effect is due to the higher inertia acquired by the plasma shell of larger mass and stronger $B_\theta$ close to the stagnation due to the rising electric current (implosion time is longer for larger $m_{0,Ar}$). $B_{z,stagnation}$ rises with $m_{0,Ar}$ as long as the time of stagnation is smaller than the rise time of the current. Therefore, for given parameters of a pulse-power system and $B_{z0}$, one can find an optimal $m_{0,Ar}$ to achieve maximum possible $B_{z,stagnation}$. For example, for the discharge parameters presented above and $B_{z0} = 0.4$ T, the optimal mass-load per length is $m_{0,Ar} \approx 200$ µg/cm. This $m_{0,Ar}$ is much bigger than the one used in the experiment, because of the reasons to be addressed in Chap. 4.

We emphasize that in the model used here, two important physical processes are omitted. The first process is the thermal pressure, which plays an important role at times close to the stagnation for implosions without initial $B_z$ or relatively small $B_{z0}$. For example, $B_{z,stagnation}$ in implosion with $B_{z0} = 0.1$ T (see Fig. 1.6) will not reach the high value of 388 T calculated by the present model because the thermal pressure (here neglected) will stop the compression already at a larger radius. The second process neglected here is the diffusion (outward) of the $B_z$-flux that can significantly effect the peak $B_z$ at stagnation. In Chap. 4 the results of the present theoretical calculations will be compared to the measured $R(t)$, $B_z(t)$, and $B_\theta(t)$.

## 1.3  Spectroscopic Diagnostics of Plasma

In the present work, we employ emission spectroscopy for the determination of the temporal and spatial evolution of the magnetic fields (axial and azimuthal $B$-fields), and the plasma parameters (electron density and temperature). Spectroscopy is a non-intrusive method for plasma diagnostics that analyses radiation emitted by the plasma. Spectroscopic data contain very rich information that is useful for determination of plasma and electromagnetic fields properties. In the following, we present features and phenomena in the emitted radiation that are used for obtaining the information:

- **Doppler shift**. Doppler shift of spectral line appears due to the relative motion of the radiating atoms (ions) towards (or away from) the observer. Measurements of Doppler shift allow for the determination of plasma velocities using: $\vec{v} \cdot \hat{n} \approx c \cdot \Delta\lambda_{Doppler}/\lambda_0$ for $v \ll c$ ($\hat{n}$ is the unit vector from the radiator to the observer, $\Delta\lambda_{Doppler} = \lambda_{measured} - \lambda_0$ is the Doppler shift of the measured spectral line-peak relative to the line-peak ($\lambda_0$) of a radiator at rest).
- **Doppler broadening**. Doppler broadening of spectral lines appears due to radiator's velocity distribution observed in the probed plasma volume or during the detection time (i.e., the resultant broadening of the line is due to the superposition of many single-ion emissions that have different Doppler shifts). Measurement of the Doppler broadening allows for the determination of the ion temperature ($T_i$) or the hydrodynamic ion velocity distribution in turbulent plasmas. If the Doppler broadening results from the ion temperature, the spectral line-profile will have a Gaussian profile with full width at half maximum $FWHM_{Gaussian}$ given by: $T_i(eV) = 1.7 \times 10^8 \cdot A \, (FWHM_{Gaussian}/\lambda_0)^2$ ($A$ is the ion mass in atomic mass unit).

- **Zeeman effect**. Zeeman splitting of the spectral line appears due to the energy splitting of atomic levels in the presence of a magnetic field. Measurement of Zeeman splitting allows for the determination of the magnetic field in plasma. For a crude estimate of the energy shift of the atomic state due to the Zeeman effect one can use $\Delta E \sim \mu_B B$ ($\mu_B$ is the Bohr magneton). For a more detailed description of the Zeeman effect see Sect. 1.3.2.
- **Stark effect**. Stark splitting of a spectral line appears due to the energy splitting of atomic levels in the presence of an electric field. Line splitting due to the Stark effect is used for the determination of "macroscopic and static" (relatively to the plasma volume and time of observation) electric fields. For hydrogen-like ions the energy shift is linearly proportional to the magnitude of the electric field, $\mid \vec{E} \mid \equiv F$, (linear Stark effect). The maximum separation between the quantum states belonging to an energy level with principal quantum number $n_p$ is given by: $\Delta E_{max}(\text{eV}) = 3ea_0 F n_p(n_p - 1)/Z \approx 1.6 \times 10^{-5} F(\text{kV/cm})n(n - 1)/Z$ (equivalently, $\Delta E_{max}(\text{cm}^{-1}) \approx 0.13 F(\text{kV/cm})n_p(n_p - 1)/Z)$, $a_0$ is the Bohr radius. For non hydrogen-like ions the energy shift is quadratic in $\mid \vec{E} \mid$ (quadratic Stark effect), and for an atomic quantum state $\mid \phi_i \rangle$ with unperturbed energy $E_i$ is given by:

$$\Delta E_i = \sum_{k \neq i} \frac{\mid \langle \phi_k \mid eFr \cos \theta \phi_i \rangle \mid^2}{E_i - E_k} \qquad (1.34)$$

Note, the atomic quantum state $\mid \phi \rangle$ is defined by the principal quantum number $n_p$, total orbital angular momentum $L$, total spin $S$, total angular momentum $J$, and projection of $J$ on the $\hat{z}$-axis, $m_J$.

To estimate the energy shift using Eq. 1.34, it is enough to include in the sum only few quantum states with closest energies for which the electric dipole matrix element is not zero (i.e., $\langle \phi_k \mid er \mid \phi_i \rangle \neq 0$). The values of the matrix elements in Eq. 1.34 can be obtained using the Einstein coefficients between the states (a convenient source for the Einstein coefficients is the on-line atomic database of NIST [50]). For a more elaborate discussion on the Stark effect see the book of H. A. Bethe and E. E. Salpeter 'Quantum mechanics of one- and two-electron atoms' [51].

- **Stark broadening**. Stark broadening of a spectral lines appears due to fluctuating electric microfields acting on the radiating ion. These microfields are generated by the moving surrounding electrons and ions. For isolated spectral lines of non hydrogen-like ions, the main contribution to the Stark broadening comes from the electric fields generated by electrons. This broadening results in a line-profile of Lorentzian distribution with $FWHM_{Lorentzian}$ that is linearly proportional to the electron density, $FWHM_{Lorentzian} = \gamma(T_e) \cdot n_e$. Usually, the $T_e$-dependence of the $FWHM_{Lorentzian}$, through the proportionality constant $\gamma$, is small in comparison to the $n_e$-dependence. Therefore, Stark broadening of spectral lines is used for electron density determination. Note, the Stark broadening of a line is usually accompanied by a Stark shift of the line-center due to the level repulsion effect, as can be seen in Eq. 1.34, if $E_k > E_i$ then contribution of the state $\mid \phi_k \rangle$ to $\Delta E_i$ is negative and vice versa, meaning that the interacting states repel each other. Since, the net energy shift of an atomic level due to the level repulsion effect can be either to higher or lower energies, the Stark shift can be either negative or positive. An additional discussion on the Stark broadening effect and its use for $n_e$ determination see Sect. 1.3.1. Much more extensive discussion on this subject can be found in the book of Griem 'Principles of plasma spectroscopy' [52].

- **Line intensity**. The intensity of a spectral line due to spontaneous emission transition between two atomic levels with upper and lower energies $E_u$ and $E_l$, respectively, is given by:

$$I\,(\mathrm{erg/s/cm}^3) = h\nu n_i\,(\mathrm{cm}^{-3})\,P_Z(n_e, T_e)\,P_{Z,u}(n_e, T_e)\,A_{ul} \qquad (1.35)$$

$h$ is the Planck constant in erg·s, $\nu$ is the transition frequency, $n_i = n_e/\bar{Z}$ is the total ion density, $P_Z$ is the probability for an ion to be in charge state $Z$, $P_{Z,u}$ is the probability of ion of charge $Z$ to be in excited atomic level $E_u$, and $A_{ul}$ is the Einstein coefficient for spontaneous transition from the quantum state of the atomic level $E_u$ to the atomic level $E_l$. Note, in $LS$ coupling approximation, the atomic level is defined by the quantum numbers $n, L, S, J$ with a degeneracy given by $g = 2J + 1$, while quantum state is not degenerate and is defined by quantum numbers $n, L, S, J, m_J$.

Since the line intensity is a function of electron density and temperature, it can be used for the determination of the plasma parameters. Usually, it is highly useful to look at the intensity ratios of different spectral line-pairs instead of the absolute intensity of each line, since some of the line-intensity ratios have dominant dependence only on a single parameter, $n_e$ or $T_e$, for a relevant range of plasma conditions. In addition, the use of the intensity ratio allows to avoid the relatively complicated process of absolute calibration of the spectroscopic system.

For example, in relatively dense plasmas, such that the electron excitation and de-excitation processes are dominant over the radiative processes (plasma is close to local thermodynamic equilibrium (LTE)), $P_{Z,u}(n_e, T_e)$ can be approximated using the Boltzmann distribution:

$$P_{Z,u}(n_e, T_e) = \mathcal{N}g_u exp(-E_u/k_B T_e) \qquad (1.36)$$

and $P_Z(n_e, T_e)$ can be approximated using Saha equation:

$$\frac{n_{Z+1}n_e}{n_Z} = \frac{P_{Z+1}n_e}{P_Z} = \frac{(2\pi m_e k_B T_e)^{3/2}}{h^3}\frac{2g_{Z+1}}{g_Z}exp(-\chi/k_B T_e) \qquad (1.37)$$

$\mathcal{N}$ is the normalization constant (1/partition function), $g_u$ is the statistical weight (degeneracy) of the atomic level $E_u$, $n_Z = n_i P_Z$ is the density of ions of charge state $Z$, $g_z$ is the statistical weight of the ground state of the ion with charge $Z$, $\chi$ is the ionization energy from the ground state of the ion of charge state $Z$. As can be seen from Eq. 1.36, in LTE plasmas the intensity ratio of spectral lines belonging to the same ion charge state depends only on $T_e$, therefore, it can be used for $T_e$ diagnostics:

$$\text{ratio} = \frac{I_A}{I_B} = \frac{h\nu_A}{h\nu_B}\frac{A_{u_A l_A}}{A_{u_B l_B}}\frac{g_{u_A}}{g_{u_B}} exp\left(-\frac{(E_{u_A} - E_{u_B})}{k_B T_e}\right) \tag{1.38}$$

By knowing $T_e$ and measuring intensities of lines belonging to different ion charge states, Saha equation (1.37) can be used for $n_e$ determination.

It is important to note that the presented discussion on the use of line intensities for plasma parameters determination assumes optically thin plasma (i.e. photon absorption is negligible). However, it is not always the case in Z-pinch plasmas, and depends very much on the selected lines (e.g., transitions to the ground state are usually optically thick), since opacity affects the escaping emission from the plasma it must be calculated for each line. More details regarding opacity calculation and how it affects the line-profile and intensity, is given at the end of the current section.

Note also, that in the above discussion, the probabilities $P_Z$ and $P_{Z,u}$ are assumed to be in steady state, therefore, they are functions of $n_e$ and $T_e$ only. However, in transient plasma case (relevant for all pulsed power experiments), the rate at which the plasma parameters changes might be faster than the time to reach steady state. In such case, the line intensity depends not only on $n_e$ and $T_e$ but also on the time history of these parameters. In order to find the true evolution of $n_e$ and $T_e$ it is necessary to fit the time evolution of different line intensities by varying the time history of $n_e(t)$ and $T_e(t)$ and using time-dependent collisional-radiative simulations [53]. Such analysis is rarely used in high-density plasmas typical for Z-pinches, due to the difficulty in obtaining reliable time history of the line-intensity (or line-profile) and large parametric space for $n_e$ and $T_e$ time-history that results in relatively large error-bars of the plasma parameters determination. For a more detailed discussion on the atomic processes in plasmas see the book of Salzman 'Atomic physics in hot plasmas' [54].

- **Continuum radiation**. Continuum radiation in plasma is emitted due to two processes:

(i) Scattering of free electrons by ions, where the electrons emit radiation during their acceleration in the electric field of ions. This process is called free-free or Bremsstrahlung radiation. For thermal electrons in a field of ions of charge state $Z$, the emitting power spectrum per unit volume is given by:

$$\epsilon^{ff}\,(\text{J/s/m}^3/\text{Hz}) = \frac{Z^2 e^6}{6\pi^2 \epsilon_0^3 m_e^2 c^3}\left(\frac{2\pi m_e}{3k_B T_e}\right)^{1/2} n_e n_i \exp(-h\nu/k_B T_e)\cdot \bar{g}_{ff}(\nu) =$$
$$= 6.8 \times 10^{-51} Z^2 n_e\,(\text{m}^{-3}) n_i\,(\text{m}^{-3}) T_e^{-1/2}\,(\text{K})\exp(-h\nu/k_B T_e)\cdot \bar{g}_{ff}(\nu) \tag{1.39}$$

in cgs units this is:

$$\epsilon^{ff} (\text{erg/s/cm}^3/\text{Hz}) = \frac{32\pi Z^2 e^6}{3m_e^2 c^3} \left(\frac{2\pi m_e}{3k_B T_e}\right)^{1/2} n_e n_i \exp(-h\nu/k_B T_e) \cdot \bar{g}_{ff}(\nu) =$$

$$= 6.8 \times 10^{-38} Z^2 n_e (\text{cm}^{-3}) n_i (\text{cm}^{-3}) T_e^{-1/2}(\text{K}) \exp(-h\nu/k_B T_e) \bar{g}_{ff}(\nu) =$$

$$= 6.3 \times 10^{-40} Z^2 n_e (\text{cm}^{-3}) n_i (\text{cm}^{-3}) T_e^{-1/2}(\text{eV}) \exp(-h\nu(\text{eV})/T_e) \cdot \bar{g}_{ff}(\nu)$$

$$(1.40)$$

and the total emitted power per unit volume is:

$$P^{ff} (\text{erg/s/cm}^3) = 1.5 \times 10^{-25} \cdot Z^2 n_e (\text{cm}^{-3}) n_i (\text{cm}^{-3}) T_e^{1/2}(\text{eV}) \bar{G}_{ff} \quad (1.41)$$

$\bar{g}_{ff}$ and $\bar{G}_{ff}$ are the velocity and velocity+frequency averaged Gaunt factor, respectively. The Gaunt factor accounts for a quantum corrections to Eqs. 1.39–1.41 which have been derived using classical physics. For the relevant plasma parameters and photon energies these two factors are in the range $1 \leq \bar{g}_{ff}, \bar{G}_{ff} \leq 2$ (see [55]).

For typical $n_e$ and $T_e$ of Z-pinch plasmas, low-Z plasmas (like hydrogen and helium) are almost fully ionized and therefore most of the energy is emitted by the Bremsstrahlung radiation. In the case of high-Z plasmas the ions still posses many bound electrons and therefore most of the radiation energy is in line spectrum.

(ii) Capture of free electrons by an ions through emission of a photon: $E_{ph} = E_{elec} - E_{bound}$, $E_{elec}$ and $E_{bound}$ are the photon energy, free electron kinetic energy and bound atomic state energy, respectively. This process is called radiative recombination or free-bound radiation. Calculation of the power spectrum due to the radiative recombination is significantly more involved than for the Bremsstrahlung radiation, since it has to be done separately for each bound atomic state to which the free electrons are recombining. The approximated expression for the emitting power spectrum per unit volume due the radiative recombination to a bound state (with $(n_p, L)$ quantum numbers) of ion with charge Z, is give by [54]:

$$\epsilon^{fb}(\text{eV/s/cm}^3/\text{eV}) = \frac{64\sqrt{\pi}}{3\sqrt{3}} \frac{e^4}{m_e^2 c^3} \frac{1}{n_p^3} \left(\frac{E_{Z-1,n_p,L}}{k_B T_e}\right)^{3/2} Z n_e N_Z \quad (1.42)$$

$$\times \exp(-h\nu/k_B T_e)(1 - P_{Z,n_p,L}) \quad \text{for } h\nu > E_{Z-1,n_p,L}$$

$E_{Z-1,n_p,L}$ is the bound energy of $|n_p, L\rangle$ atomic state of ion with $Z - 1$ charge state, $P_{Z,n_p,L}$ is the population probability of the $|n_p, L\rangle$ atomic state in ion with charge state Z, $N_Z$ is the particle density of ions with charge state Z.

Continuum radiation can be used for the determination of both, $n_e$ and $T_e$. For example, in plasmas for which a wide spectral range of the continuum spectrum ($\epsilon^{ff}(\nu)$) can be measured accurately, $T_e$ can be calculated from the slope of the linear fit to $ln(\epsilon^{ff}(\nu))$ versus $\nu$ plot. Subsequently, $n_e$ can be determined by measuring the absolute continuum emission intensity within some spectral range, and using Eq. 1.39. It is important to note that for an accurate determination of $n_e$, $\bar{Z}$ should be known.

We now discuss two issues that might complicate the spectroscopic data analysis and may lead to large errors in the determination of plasma and electromagnetic fields properties if not treated properly. The first is the re-absorption of photons in the plasma (opacity effect). The second is the spectrum that contains emission from plasmas of very different properties, either due to the integration along line-of-sight or due to the low spatial or temporal resolutions.

(i) **Optical thickness of lines**. Here, we discuss only optical thickness of spectral lines, since for the relevant plasma parameters the absorption of continuum is negligible. The optical thickness $\tau(\nu)$ is a dimensionless parameter defined by:

$$\tau(\nu) = \int_{s_1}^{s_2} \alpha_{abs}(\nu, s)ds \qquad (1.43)$$

where $\alpha_{abs}(\nu, s)$ is the absorption coefficient (1/mean-free path) of a photon with frequency $\nu$, and the integration is along the line-of-sight inside the plasma.

The calculation of the optical thickness for the spectral lines used for the diagnostics is important, since $\tau(\nu)$ effects both, the measured intensity of the lines and their profiles. Usually, the optical thickness is calculated at the peak of the spectral line assuming uniform plasma of characteristic length $L_{pl}$:

$$\tau_{0,lu} = \frac{1}{8\pi} \lambda^2 \frac{g_u}{g_l} A_{ul} n_l \left(1 - \frac{g_l n_u}{g_u n_l}\right) \phi_{ul}(\nu = \nu_0) L_{pl} = \qquad (1.44)$$

$$= 0.0265 \cdot f_{lu} n_l (\text{cm}^{-3}) \left(1 - \frac{g_l n_u}{g_u n_l}\right) \phi_{ul}(\nu = \nu_0) L_{pl}(\text{cm})$$

$\lambda(\text{cm})$ is the wavelength of the spectral line, $n_u(\text{cm}^{-3}) = n_i P_Z P_{Z,l}$ and $n_l(\text{cm}^{-3}) = n_i P_Z P_{Z,l}$ are the densities of the upper and lower atomic levels, respectively, $\phi_{ul}(\nu = \nu_0)$ is the value of the line intensity distribution (i.e line-profile normalized to unit area) at the line-peak, $f_{lu}$ is the absorption oscillator strength:

$$f_{lu} = \frac{m_e c}{8\pi e^2} \lambda^2 (\text{cm}) \frac{g_u}{g_l} A_{ul} = 1.5 \cdot \lambda^2 (\text{cm}) \frac{g_u}{g_l} A_{ul} \qquad \text{cgs units} \qquad (1.45)$$

In many cases it is reasonable to make the approximation: $1 - g_l n_u / g_u n_i \approx 1$. Using this approximation and Eq. 1.44, convenient formulas for estimation of $\tau_{0,lu}$ can be derived:

$$\tau_{0,lu}^G \approx 8.3 \times 10^{-21} \cdot n_l (\text{cm}^{-3}) f_{lu} \frac{\lambda^2(\text{Å})}{\Delta\lambda_G(\text{Å})} L_{pl}(\text{cm}) \qquad \text{for Gaussian line-profile}$$

$$(1.46)$$

$$\tau_{0,lu}^L \approx 5.63 \times 10^{-21} \cdot n_l (\text{cm}^{-3}) f_{lu} \frac{\lambda^2(\text{Å})}{\Delta\lambda_L(\text{Å})} L_{pl}(\text{cm}) \qquad \text{for Lorentzian line-profile}$$

$$(1.47)$$

$\Delta\lambda_G(\text{Å})$ and $\Delta\lambda_L(\text{Å})$ are the full width at half maximum of the Gaussian and Lorentzian distributions, respectively. A spectral line is called optically thick if $\tau_{0,lu} > 1$ and optically thin if $\tau_{0,lu} < 1$.

The effect of opacity on the spectral line-profile and intensity is given by:

$$\phi_{ul}(\nu) = \frac{1}{L_{pl}} \int_0^{L_{pl}} \phi_{0,ul}(\nu) exp\left(-\frac{\tau_{0,lu}}{L_{pl}}\phi_{0,ul}(\nu)s\right) ds = \frac{1}{\tau_{0,lu}}(1 - exp(-\tau_{0,lu}\phi_{0,ul}(\nu)))$$

$$(1.48)$$

$\phi_{0,ul}$ is the line-profile without opacity effect (i.e., for $\tau_{0,lu} = 0$) normalized to $\phi_{0,ul}(\nu = \nu_0) = 1$. It is important to note that opacity affects the line intensity by two processes. The first is a change of the level populations due to resonance photo-absorption or photoionization, and the second is a reduction of the line-intensity due to the absorption of the photons along the line-of-sight. Usually, for plasma parameters relevant to Z-pinches, it is enough to consider only the second process (as done by Eq. 1.48), since the first process has a negligible effect on the line-intensity due to the high electron deexcitation rate (relative to the radiative decay rate) of the upper level.

(ii) **Spectrum containing emission from regions with different plasma properties**. Figure 1.8 presents a schematic description of the basic elements of an imaging setup that contains a source (plasma), imaging optics (lens, mirror), and a detector (camera, photodiode, photomultiplier, etc). The enlarged part of the optical setup presents a cross section of the light-collection cone in a plasma. For a given plasma source, the volume of the light-collection cone is determined by the working $f$-number and the diffraction limit or detector size (pixel size in a camera, optical fiber diameter, or photodiode size). In general, the diameter of this cone at the outer edge of the plasma (in the direction perpendicular to the line-of-sight) determines the spatial resolution of the imaging setup since an radiator emitting within the plasma volume defined by this cone will contribute to the signal measured by the elementary detection cell (e.g. camera pixel, single

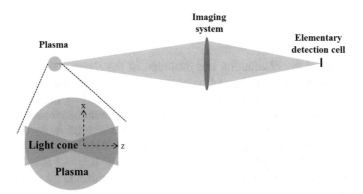

**Fig. 1.8** Schematic description of a basic elements of an imaging setup. Enlarged image shows the light collection cone in a plasma

photodiode). If the spatial resolution $d_{res}$ is smaller than the scale over which the plasma parameters change significantly (for example $\frac{1}{n_e}\frac{dn_e}{dx} \cdot d_{res} \gtrsim 1$), then the spectrum will contain emissions from plasmas with significantly different properties. Such integrated spectrum causes complications in the analysis of the line-profiles and line-intensities. A good example of the effect of the spatial integration on the line-profiles is discussed in [56]. Similarly, if the time-resolution of the measurements is lower than the rate of the change of the plasma parameters then the time-integrated spectrum will contain emissions from significantly different plasmas. A good example of the effect of the temporal integration on the line-profiles is discussed in [57]. In addition to the spatial integration perpendicular to the line-of-sight due to the finite optical resolution, there is spatial integration along line-of-sight, as can be seen in Fig. 1.8. However, for plasmas possessing cylindrical symmetry, it is possible to obtain information on the radial distributions from the line-integrated data by applying inverse Abel transform to the spectral images [58, 59]. Otherwise, a local doping by atomic species different from the surrounding plasma can be very useful to obtain local information without the complication arising due to the integration along line-of-sight, as was done in [60, 61].

We emphasize that in the present work, much effort is invested in the simultaneous determination of the various parameters, either by finding spectral regions, or even specific transitions that their analysis provide such information. For example, a Zeeman split pattern that gives the magnetic field, whereas the width of each Zeeman component gives the electron density, as was done in [56]. The simultaneous measurement of the plasma and magnetic field parameters is very important in such pulsed power experiments due to the irreproducibility in the Z-pinch evolution as a result of the development of different instabilities during the implosion and stagnation. The details of the specific transitions utilized for each measurement are presented in the following sections.

### 1.3.1 Determination of Plasma Parameters

Determination of the distribution and evolution of the electron density and temperature is essential for the calculation of the different plasma properties (thermal pressure, electrical conductivity, plasma frequency, etc.) and for understanding the physical processes occurring in Z pinches. Here, $n_e$ is determined from Stark broadening of Ar III $(^4S)4s\ ^5S_2 - (^4S)4p\ ^5P_2$ transition at $\lambda = 3302$ Å, Ar III $(^2D)4s\ ^3D_3 - (^2D)4p\ ^3P_2$ transition at $\lambda = 2884$ Å, and Ar IV $(^3P)4s\ ^2P_{3/2} - (^3P)4p\ ^2D_{5/2}$ transition at $\lambda = 2913$ Å. For the relevant $n_e$ range ($5 \times 10^{17} - 3 \times 10^{18}$ cm$^{-3}$) at $T_e \approx 4$ eV, these lines are isolated and their profile is dominated by the Lorentzian distribution due to the Stark broadening. The calculated Stark broadening of these transitions for $T_e \approx 4$ eV is [62–65]:

**Table 1.5** Optical thickness of Ar III $(^4S)4s\,^5S_2 - (^4S)4p\,^5P_2$ transition at $\lambda = 3302$ Å, Ar III $(^2D)4s\,^3D_3 - (^2D)4p\,^3P_2$ transition at $\lambda = 2884$ Å, and Ar IV $(^3P)4s\,^2P_{3/2} - (^3P)4p\,^2D_{5/2}$ transition at $\lambda = 2913$ Å, for different $n_e$ and $T_e$, and for plasma length = 0.5 cm

| $n_e$ (cm$^{-3}$) | $10^{17}$ | $10^{17}$ | $10^{18}$ | $10^{18}$ | $5 \times 10^{18}$ | $5 \times 10^{18}$ |
|---|---|---|---|---|---|---|
| $T_e$ (eV) | 4 | 5 | 4 | 5 | 4 | 5 |
| $\lambda = 2884$ Å | 0.09 | 0.04 | 0.4 | 0.16 | 0.94 | 0.56 |
| $\lambda = 3302$ Å | 0.2 | 0.09 | 0.88 | 0.3 | 1.9 | 1 |
| $\lambda = 2913$ Å | 0.09 | 0.33 | 0.2 | 1 | 0.13 | 1.1 |

$$\Delta\lambda_{Stark}^{3302}(\text{Å}) = 1.4 \times 10^{-18} \cdot n_e(\text{cm}^{-3}) \tag{1.49}$$

$$\Delta\lambda_{Stark}^{2884}(\text{Å}) = 1.1 \times 10^{-18} \cdot n_e(\text{cm}^{-3}) \tag{1.50}$$

$$\Delta\lambda_{Stark}^{2913}(\text{Å}) = 0.7 \times 10^{-18} \cdot n_e(\text{cm}^{-3}) \tag{1.51}$$

$\Delta\lambda_{Stark}$ is the full width at half maximum of the Lorentzian distribution. We note that the uncertainty of the proportionality constant $\gamma$ for the above transitions is $\sim$40% and the values given in Eqs. 1.49–1.51 are the mean of the values found in [62–65].

Table 1.5 gives the calculated $\tau_{0,lu}$ using Eq. 1.44 for the discussed above transitions in the relevant plasma parameters. $n_l$ and $n_u$ were calculated using collisional-radiative (CR) model assuming steady state using the code NOMAD [53], $f_{lu}$ values are taken from the NIST online database [50], Stark broadening for each line is calculated using Eqs. 1.49–1.51, and the plasma length is twice the width of the plasma-shell observed in the experiments, based on spectral images. It is seen from Table 1.5 that for most of the considered plasma parameters $\tau_{0,lu} \leq 1$, i.e. the lines are optically thin. However, using Eq. 1.48 it can be shown that in the case of $\tau_{0,lu} = 1$, Lorentzian line-profile is broadened by $\sim$30%. Therefore, for some relevant plasma parameters, the opacity broadening should be considered when the Stark broadening is used for $n_e$ determination.

In the present work, the electron temperature diagnostics is based on line-intensity ratios. We use the ratio of two Ar III lines, $(^4S)4s\,^5S_2 - (^4S)4p\,^5P_2$ transition at $\lambda = 3311$ Å and $(^2D)4s\,^3D_3 - (^2D)4p\,^3F_1$ transition at $\lambda = 3336$ Å which are close in wavelength and can be detected simultaneously by the same spectrometer. The advantages of using the intensity ratios of transitions belonging to the same ion are fast establishment of a steady state ratio ($<$1 ns, for the typical plasma parameters of the experiment) and weak $n_e$ sensitivity. Figure 1.9 shows the line-intensity ratio as a function of time for three different electron densities, $n_e = 10^{17}$, $10^{18}$ and $5 \times 10^{18}$ cm$^{-3}$. The ratios are calculated using population probabilities of the atomic levels obtained by the CR code NOMAD [53], and the Einstein coefficients taken from the NIST online database [50]. In the figure, stars represent line-intensity ratio for plasma in LTE, i.e. they are obtained using Eq. 1.38. It can be seen from Fig. 1.9 that already for $n_e = 10^{17}$ cm$^{-3}$ the ratios obtained in CR steady state are close to the ratios in LTE, making these lines a good choice for $T_e$ determination also when detailed CR code is unavailable.

**Fig. 1.9** Intensity ratio of Ar III $(^4S)4s\,^5S_2 - (^4S)4p\,^5P_2$ transition at $\lambda = 3311$ Å to Ar III $(^2D)4s\,^3D_3 - (^2D)4p\,^3F_4$ transition at $\lambda = 3336$ Å as a function of $T_e$ for different $n_e$. The black starts show the intensity ratio for plasma in local thermodynamic equilibrium (i.e Boltzmann distribution for atomic levels population of Ar III ion)

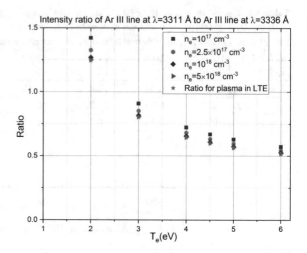

We also use line-intensity ratio of transitions belonging to different ionization states. The advantage of this method is a high sensitivity to $T_e$. Here, we employ Ar III $(^2D)4s\,^3D_3 - (^2D)4p\,^3P_2$ transition at $\lambda = 2884$ Å, and Ar IV $(^3P)4s\,^2P_{3/2} - (^3P)4p\,^2D_{5/2}$ transition at $\lambda = 2913$ Å for the $T_e$ determination of plasma filaments that form during the implosion (see Sect. 3.3.2). Figure 1.10 shows the line-intensity ratio as a function of electron temperature for three different electron densities, $n_e = 10^{17}$, $10^{18}$ and $5 \times 10^{18}$ cm$^{-3}$. Also here, the ratios are calculated using population probabilities of the atomic levels obtained by NOMAD [53], where the Einstein coefficients are taken from the NIST online database [50]. It is seen from Fig. 1.10 that the ratio is very sensitive to $T_e$, but also has non negligible dependence on $n_e$. However, since here the $n_e$ is determined employing Stark broadening, also line-pairs that are sensitive to $n_e$ can be used. In Fig. 1.10 the solid and hollow symbols represent line-intensity ratios for CR steady state and LTE, respectively. We see here that for the selected transitions, LTE ratio can be used for $n_e > 10^{18}$ cm$^{-3}$ by applying Eqs. 1.36 and 1.37.

Important point that has to be considered when the line-ratio of transitions belonging to different ion charge states are used is the time required to establish steady state. Figure 1.11 shows the line-intensity ratio of Ar III $(^2D)4s\,^3D_3 - (^2D)4p\,^3P_2$ transition at $\lambda = 2884$ Å to Ar IV $(^3P)4s\,^2P_{3/2} - (^3P)4p\,^2D_{5/2}$ transition at $\lambda = 2913$ Å as a function of time for the typical plasma parameters in the present experiment. The evolution of this ratio is calculated using CR code NOMAD with the initial condition that all the ions are Ar III and in the ground state. It is seen from Fig. 1.11 that the CR steady state is reached on the time scales of $\sim$20 ns and $\sim$70 ns for $n_e = 10^{18}$ cm$^{-3}$, $T_e = 5$ eV and $n_e = 10^{18}$ cm$^{-3}$, $T_e = 4$ eV, respectively. Since, the plasma implosion occurs on the time-scale of $\sim$1 $\mu$s the line-ratio in CR steady state of Ar III transition at $\lambda = 2884$ Å and Ar IV transition at $\lambda = 2913$ Å can be used.

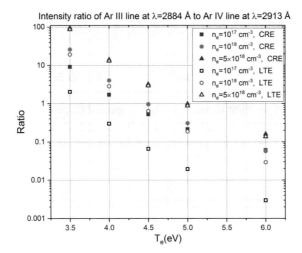

**Fig. 1.10**  Intensity ratio of Ar III $(^2D)4s\,^3D_3 - (^2D)4p\,^3P_2$ transition at $\lambda = 2884$ Å to Ar IV $(^3P)4s\,^2P_{3/2} - (^3P)4p\,^2D_{5/2}$ transition at $\lambda = 2913$ Å as a function of $T_e$ for different $n_e$. The filled squares, circles, and triangles show the intensity ratio for plasma in collisional-radiative steady state, while the hollow squares, circles, and triangles show the intensity ratio for plasma in local thermodynamic equilibrium

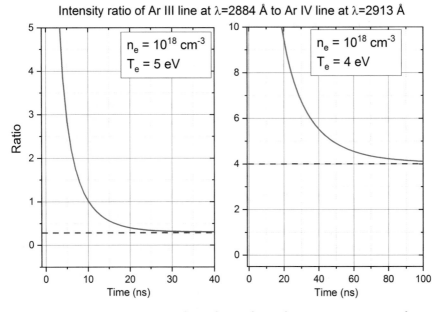

**Fig. 1.11**  Line-intensity ratio of Ar III $(^2D)4s\,^3D_3 - (^2D)4p\,^3P_2$ transition at $\lambda = 2884$ Å to Ar IV $(^3P)4s\,^2P_{3/2} - (^3P)4p\,^2D_{5/2}$ transition at $\lambda = 2913$ Å as a function of time for $n_e = 10^{18}$ cm$^{-3}$ and $T_e = 4$ and $5$ eV. The initial condition is all the ions at $t = 0$ are in the ground state of Ar III. The dashed blue line represents the intensity ratio for plasma in collisional-radiative steady state

### 1.3.2  Determination of Magnetic Field

In the present study, Zeeman effect is used to measure magnetic fields in the plasma. The Zeeman effect results in the splitting of atomic or ionic degenerate energy levels due to the interaction with an external magnetic field. If the energy shifts due to the Zeeman effect are small in comparison to the spin-orbit interaction (this condition is satisfied in all of our measurements) then the energy splitting of the considered atomic (or ionic) level is calculated using the approximation:

$$\Delta E = g_{LSJ} \mu_B m B, \tag{1.52}$$

where $m$ is the projection of the total angular momentum $J$ of the given state along the direction of the magnetic field $B$, $\mu_B$ is the Bohr magneton, and $g_{LSJ}$ is the Lande $g$-factor, given by:

$$g_{LSJ} = 1 + \frac{J(J+1) + S(S+1) - L(L-1)}{2J(J+1)}, \tag{1.53}$$

where $S$ and $L$ are, respectively, the total spin and the orbital angular momentum of the radiator. The relative intensities of the various Zeeman components for a transition between levels with total angular momentum $J$ and $J'$ are given by the Wigner 3-j symbols (see for example [66]):

$$I \sim \begin{pmatrix} J & 1 & J' \\ -m & m-m' & m' \end{pmatrix}^2 \tag{1.54}$$

For relatively simple atomic configurations, like single optical electron above a filled atomic shell, the calculation of the exact Zeeman-split pattern is performed by diagonalizing the following Hamiltonian:

$$H = H_0 + \xi \vec{L} \cdot \vec{S} + (L_z + 2S_z)\mu_B B \tag{1.55}$$

consisting of the zero-order Hamiltonian, the $LS$ interaction term, and the magnetic field interaction term, respectively. $L_z$ and $S_z$ are, respectively, the projection of the total orbital angular momentum $L$ and total spin $S$ of the given state along the direction of the magnetic field $B$. The coefficient $\xi$ is found using published energy levels ([50]).

An example of the $B$-field induced energy splitting of the Al III $4s\ ^2S_{1/2}$ and $4p\ ^2P_{1/2}$ atomic levels is presented in Fig. 1.12. The transition between these two atomic levels was used in the present work for axial magnetic field ($B_z$) determination. The arrows show the electric dipole allowed transitions between states with different $m$ values. The red arrows correspond to the transitions with $\Delta m = m_{upper} - m_{lower} = 0$, the blue arrows correspond to the transitions with $\Delta m = \pm 1$. Figure 1.13 shows the simulation of the Al III $4s\ ^2S_{1/2} - 4p\ ^2P_{3/2}$ transition line

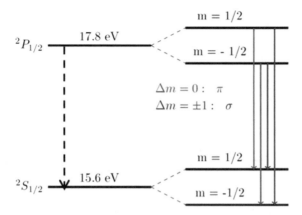

**Fig. 1.12** Energy diagram of the Al III $4s\ ^2S_{1/2} - 4p\ ^2P_{1/2}$ transition in the presence of a magnetic field. The dashed arrow represents the unperturbed transition

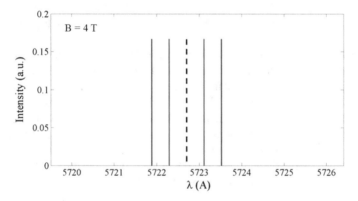

**Fig. 1.13** The simulated Al III $4s\ ^2S_{1/2} - 4p\ ^2P_{1/2}$ Zeeman-split transitions for $B = 4$ T. The black dashed line represents the spectral line position without the external magnetic field. Line of sight is perpendicular to the $B$-field direction

splitting due to the Zeeman effect for $B = 4$ T and line of sight perpendicular to the $B$-field direction. In relatively dense plasmas ($n_e > 10^{17}$ cm$^{-3}$), the measurement of a magnetic field in the range of a few Tesla, using the Zeeman effect is challenging due to the large Stark broadening of the transition line that smears out the split pattern (see Fig. 1.14). Stark broadening of lineshape arises from fluctuating electric fields at the radiator location, generated by the moving plasma charged particles (electrons and ions). For non-hydrogenlike ions and isolated transitions, the Stark broadening is proportional to $n_e$ and has a Lorentzian shape. Figure 1.14 demonstrates the effect of an electron density of $n_e = 5 \times 10^{17}$ cm$^{-3}$ on the Zeeman pattern calculated for Al III $4s\ ^2S_{1/2} - 4p\ ^2P_{1/2}$ transition at $\lambda = 5722.7$ Å for $B = 4$ T and a line of

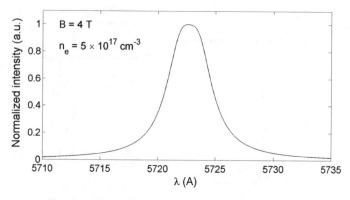

**Fig. 1.14** Simulated Al III $4s\ {}^2S_{1/2} - 4p\ {}^2P_{1/2}$ transition lineshape for $B = 4$ T and Stark broadening corresponding to $n_e = 5 \times 10^{17}$ cm$^{-3}$. Line of sight is perpendicular to the $B$-field direction

sight perpendicular to the magnetic field direction. Similar difficulty may arise in a hot or turbulent plasmas where the Doppler broadening smears out the Zeeman-split pattern.

In observations perpendicular to the magnetic field direction, the photons emitted in the $\Delta m = 0$ transitions (red arrows in Fig. 1.12) are linearly polarized along the $B$-field direction (defined as $\pi$ polarization), whereas the photons from the $\Delta m = \pm 1$ transitions (blue arrows in Fig. 1.12) are linearly polarized perpendicular to the $B$-field direction (defined as $\sigma$ polarization). Since the $\Delta m = 0$ and $\Delta m = \pm 1$ transitions have different splitting, and the Stark, Doppler, and instrumental broadenings are the same for each splitting component, the comparison of the lineshapes of the two different polarizations allows for the $B$-field determination [67], even if the Zeeman splitting is not resolved. Figure 1.15 shows the comparison of the two simulated lineshapes for $\pi$ (red) and $\sigma$ (blue) polarizations with the same plasma and

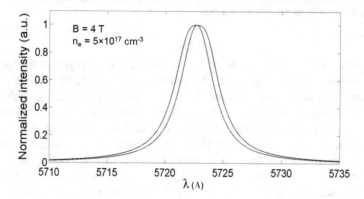

**Fig. 1.15** The red and blue solid lines are simulated Al III $4s\ {}^2S_{1/2} - 4p\ {}^2P_{1/2}$ transition lineshapes of $\pi$ and $\sigma$ polarizations, respectively, for $B = 4$ T and $n_e = 5 \times 10^{17}$ cm$^{-3}$

**Fig. 1.16** Energy diagram of the Ar III $(^4S)4s\,^5S_2 - (^4S)4p\,^5P_2$ transition in the presence of a magnetic field

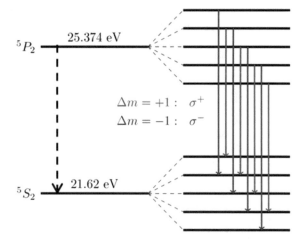

magnetic field parameters ($B = 4$ T, $n_e = 5 \times 10^{17}$ cm$^{-3}$). The difference in the lineshape widths of the two polarizations is used for the axial magnetic field determination.

For the azimuthal magnetic field ($B_\theta$) determination the emission from the imploding argon plasma was used. In Fig. 1.16, an example of energy splitting due to the Zeeman effect is given for Ar III $(^4S)4s\,^5S_2$ and $(^4S)4p\,^5P_2$ atomic levels. In the present work, the transition between these two atomic levels was used for the measurements of $B_\theta$ that is generated by the current carried by the plasma. The arrows show the electric dipole allowed transitions between states with $\Delta m = \pm 1$. These are the transitions observed when the viewing direction is parallel to the magnetic field. The resultant line splitting for $B = 4$ T is shown in Fig. 1.17. For electron densities $n_e > 10^{18}$ cm$^{-3}$ and $B$-fields of several Tesla, which are the typical values of $n_e$ and $B_\theta$ in our imploding plasma shell experiments, the Zeeman pattern of the transition is unresolvable as can be seen in Fig. 1.18. Therefore, similarly to the $B_z$ measurements, also for the $B_\theta$ determination we employed a technique that is based on the polarization properties of the Zeeman line splitting.

In observation parallel to the magnetic field direction, photons emitted in $\Delta m = 1$ transitions (red arrows in Fig. 1.16) are circularly polarized and are defined as $\sigma^+$ polarization. Photons emitted in $\Delta m = -1$ transitions (blue arrows Fig. 1.16) are also circularly polarized, but in the opposite direction than the $\sigma^+$ photons, and are defined as $\sigma^-$ polarization. Since the $\Delta m = 1$ and $\Delta m = -1$ transitions are shifted in different direction, the measurement of the relative spectral shift between $\sigma^+$ and $\sigma^-$ allows for the $B$-field determination. Figure 1.19 presents two lineshapes which are simulated for the same plasma and magnetic field parameters ($B = 4$ T, $n_e = 2 \times 10^{18}$ cm$^{-3}$), the red profile represents the lineshape of $\sigma^+$-polarization and the blue profile represents the lineshape of $\sigma^-$-polarization. The relative shift between the lineshapes of the two polarizations was used for the $B_\theta$ determination.

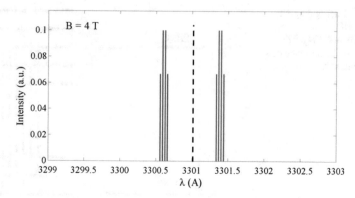

**Fig. 1.17** Calculated Ar III $(^4S)4s$ $^5S_2 - (^4S)4p$ $^5P_2$ transition line splitting for $B_\theta = 4$ T. The black dashed line represents the spectral position of the transition without external magnetic field. Line of sight is parallel to the $B$-field direction

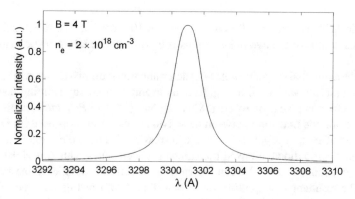

**Fig. 1.18** Simulated Ar III $(^4S)4s$ $^5S_2 - (^4S)4p$ $^5P_2$ transition lineshape for $B = 4$ T and Stark broadening corresponding to $n_e = 2 \times 10^{18}$ cm$^{-3}$. Line of sight is parallel to the $B$-field direction

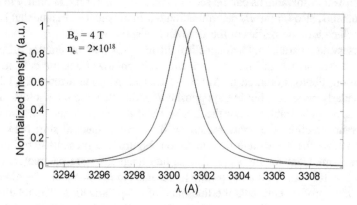

**Fig. 1.19** The red and blue solid lines are simulated Ar III $(^4S)4s$ $^5S_2 - (^4S)4p$ $^5P_2$ transition lineshapes of $\sigma^+$ and $\sigma^-$ polarizations, respectively, for $B = 4$ T and $n_e = 2 \times 10^{18}$ cm$^{-3}$

# References

1. Birn J, Schindler K (2002) Thin current sheets in the magnetotail and the loss of equilibrium. J Geophys Res 107:1117
2. Stenflo JO (2010) Distribution function for magnetic fields on the quiet sun. Astron Astrophys 517:A37
3. Kroupp E, Osin D, Starobinets A, Fisher V, Bernshtam V, Maron Y, Uschmann I, Förster E, Fisher A, Deeney C (2007) Ion-kinetic-energy measurements and energy balance in a z-pinch plasma at stagnation. Phys Rev Lett 98(11):115001
4. McGuire KM et al (1995) Review of deuterium-tritium resulst from the tokamak fusion test reactor. Phys Plasmas 2:2176
5. Martinez-Sanchez M, Pollar JE (1998) Spacecraft electric propulsion-an overview. J Propuls Power 14:688
6. Matsumoto M, Murakami T, Okuno Y (2010) Two-dimensional numerical simulation on the mhd flow behavior in a pulse-detonation-driven mhd electrical power generator. IEEJ Trans Electr Electron Eng 5:422
7. Slutz SA et al (2010) Pulsed-power-driven cylindrical liner implosions of laser preheated fuel magnetized with an axial fields. Phys Plasmas 17(5):
8. Spielman RB, Deeney C, Chandler GA, Douglas MR, Fehl DL, Matzen MK, McDaniel DH, Nash TJ, Porter JL, Sanford TWL, Seamen JF, Stygar WA, Struve KW, Breeze SP, McGurn JS, Torres JA, Zagar DM, Gilliland TL, Jobe DO, McKenney JL, Mock RC, Vargas M, Wagoner T, Peterson DL (1998) Tungsten wire-array z-pinch experiments at 200 tw and 2 mj. Phys Plasmas 5(5):2105–2111
9. Bellan PM et al (2009) Astrophysical jets: observations, numerical simulations, and laboratory experiments. Phys Plasmas 16:041005
10. Lebedev SV, Ciardi A, Ampleford D, Bland SN, Bott SC, Chittenden JP, Hall G, Rapley J, Frank A, Blackman EG, Lery T (2005) Magnetic tower outflows from a radial wire array z-pinch. Mon Not R Astron Soc 361:97
11. Lebedev SV (2007) High energy density laboratory astrophysics. Springer, Netherlands
12. Felber FS, Wessel FJ, Wild NC, Rahman HU, Fisher A, Fowler CM, Liberman MA, Velikovich AL (1988) Ultrahigh magnetic fields produced in a gas-puff z pinch. J Appl Phys 64(8):3831–3844
13. Felber FS, Malley MM, Wessel FJ, Matzen MK, Palmer MA, Spielman RB, Liberman MA, Velikovich AL (1988) Compression of ultrahigh magnetic fields in a gas-puff z pinch. Phys Fluids 31(7):2053–2056
14. Appartaim RK, Dangor AE (1998) Large magnetic fields generated by z-pinch flux compression. J Appl Phys 84(8):4170–4175
15. Zukakishvili GG, Mitrofanov KN, Grabovskii EV, Oleinik GM (2005) Measurements of the axial magnetic field during the implosion of wire arrays in the angara-5-1 facility. Plasma Phys Rep 31:652–664
16. Gotchev OV et al (2009) Laser-driven magnetic-flux compression in high-energy-density plasmas. Phys Rev Lett 103(4):215004
17. Sorokin SA (2010) High-magnetic-field generation by magnetic flux compression in imploding plasma liners. Trans Plasma Sci 38(8):1723–1725
18. Felber FS, Liberman MA, Velikovich AL (1988) Magnetic flux compression by dynamic plasmas. I. subsonic self-similar compression of a magnetized plasma-filled liner. Phys Fluids 31(12):3675–3682
19. Felber FS, Liberman MA, Velikovich AL (1988) Magnetic flux compression by dynamic plasmas. II. supersonic self-similar solutions for magnetic cumulation. Phys Fluids 31(12):3683–3689

20. Budko AB, Felber FS, Kleev AI, Liberman MA, Velikovich AL (1989) Stability analysis of dynamic z pinches and theta pinches. Phys Fluids B Plasma Phys 1(3):598–607
21. Chaikovsky SA et al (2003) The K-shell radiation of a double gas puff z-pinch with an axial magnetic field. Laser Part Beams 21:255–264
22. Shishlov AV, Baksht RB, Chaikovsky SA, Fedunin AV, Fursov FI, Kokshenev VA, Kurmaev NE, Labetsky AYu, Oreshkin VI, Ratakhin NA, Russkikh AG, Shlykhtun SV (2006) Formation of tight plasma pinches and generation of high-power soft x-ray radiation pulses in fast z-pinch implosions. Laser Phys 16(1):183–193
23. Gomez MR, Slutz SA, Sefkow AB, Sinars DB, Hahn KD, Hansen SB, Harding EC, Knapp PF, Schmit PF, Jennings CA, Awe TJ, Geissel M, Rovang DC, Chandler GA, Cooper GW, Cuneo ME, Harvey-Thompson AJ, Herrmann MC, Hess MH, Johns O, Lamppa DC, Martin MR, McBride RD, Peterson KJ, Porter JL, Robertson GK, Rochau GA, Ruiz CL, Savage ME, Smith IC, Stygar WA, Vesey RA (2014) Experimental demonstration of fusion-relevant conditions in magnetized liner inertial fusion. Phys Rev Lett 113:155003
24. Mikitchuk D, Stollberg C, Doron R, Kroupp E, Maron Y, Strauss HR, Velikovich AL, Giuliani JL (2014) Mitigation of instabilities in a z-pinch plasma by a preembedded axial magnetic field. Trans Plasma Sci 42(10):2524–2525
25. Mikitchuk D, Cvejic M, Doron R, Kroupp E, Stollberg C, Maron Y, Velikovich AL, Ouart ND, Giuliani JL, Mehlhorn TA, Yu EP, Fruchtman A (2019) Effects of a preembedded axial magnetic field on the current distribution in a Z-pinch implosion. Phys Rev Lett 122:045001
26. Rousskikh AG, Zhigalin AS, Oreshkin VI, Frolova V, Velikovich AL, Yushkov GYu, Baksht RB (2016) Effect of the axial magnetic field on a metallic gas-puff pinch implosion. Phys Plasmas 23(6)
27. Rousskikh AG, Zhigalin AS, Oreshkin VI, Baksht RB (2017) Measuring the compression velocity of a z pinch in an axial magnetic field. Phys Plasmas 24(6):063519
28. Yager-Elorriaga DA, Lau YY, Zhang P, Campbell PC, Steiner AM, Jordan NM, McBride RD, Gilgenbach RM (2018) Evolution of sausage and helical modes in magnetized thin-foil cylindrical liners driven by a z-pinch. Phys Plasmas 25(5):056307
29. Yager-Elorriaga DA, Zhang P, Steiner AM, Jordan NM, Campbell PC, Lau YY, Gilgenbach RM (2016) Discrete helical modes in imploding and exploding cylindrical, magnetized liners. Phys Plasmas 23(12):124502
30. Qi N, de Grouchy P, Schrafel PC, Atoyan L, Potter WM, Cahill AD, Gourdain P-A, Greenly JB, Hammer DA, Hoyt CL, Kusse BR, Pikuz SA, Shelkovenko TA (2014) Gas puff z-pinch implosions with external Bz field on cobra. AIP conference proceedings 1639(1):51–54
31. Qi N, Rocco SV, Banasek JT, Atoyan L, Byvank T, Potter WM, Greenly JB, Hammer DA, Kusse BR (2018) Ar gas-puff z-pinches with applied Bz on cobra at cornell university. In: International conference on plasma sciences
32. Beg FN, Conti F, Valenzuela J, Aybar N, Narkis J, Ruskov E, Covington A, Rahman H (2018) Effect of an axial magnetic field on stabilization of magneto-rayleigh-taylor instability in gas-puff z-pinches. In: 60th annual meeting of the APS division of plasma physics
33. Conti F, Valenzuela J, Ross MP, Narkis J, Beg F (2018) Stability measurements of a staged z-pinch with applied axial magnetic field. In: The 45th IEEE international conference on plasma science
34. Rahman H, Ruskov E, Ney P, Narkis J, Conti F, Valenzuela J, Ross MP, Beg F, Anderson A, Dutra E, Covington A (2018) Staged z-pinch experiments on zebra and simulations using different gas shells. In: The 45th IEEE international conference on plasma science
35. Giuliani JL, Commisso RJ (2015) A review of the gas-puff z-pinch as an x-ray and neutron source. IEEE Trans Plasma Sci 43(8):2385–2453
36. Haines MG (2011) A review of the dense z-pinch. Plasma Phys Control Fusion 53(9):093001
37. Liberman MA, Spielman RB, Toor A, Groot JS (1999) Physics of high-density z-pinch plasmas. Springer, New York
38. Ryutov DD et al (2000) The physics of fast z pinches. Rev Mod Phys 72:167
39. Spielman RB, De Groot JS (2001) Z pinches-a historical view. Laser Part Beams 19(4):509–525

40. Dangor AE (1986) High density z-pinches. Plasma Phys Control Fusion 28(12B):1931
41. Kroupp E, Osin D, Starobinets A, Fisher V, Bernshtam V, Weingarten L, Maron Y, Uschmann I, Förster E, Fisher A, Cuneo ME, Deeney C, Giuliani JL (2011) Ion temperature and hydrodynamic-energy measurements in a z-pinch plasma at stagnation. Phys Rev Lett 107:105001
42. Lash J (2015) Capability advances at the sandia z machine
43. Sinars DB, Slutz SA, Herrmann MC, McBride RD, Cuneo ME, Peterson KJ, Vesey RA, Nakhleh C, Blue BE, Killebrew K, Schroen D, Tomlinson K, Edens AD, Lopez MR, Smith IC, Shores J, Bigman V, Bennett GR, Atherton BW, Savage M, Stygar WA, Leifeste GT, Porter JL (2010) Measurements of magneto-rayleigh-taylor instability growth during the implosion of initially solid al tubes driven by the 20-MA, 100-ns z facility. Phys Rev Lett 105:185001
44. Velikovich AL, Cochran FL, Davis J, Chong YK (1998) Stabilized radiative z-pinch loads with tailored density profiles. Phys Plasmas 5(9):3377–3388
45. Sze H, Banister J, Failor BH, Levine JS, Qi N, Velikovich AL, Davis J, Lojewski D, Sincerny P (2005) Efficient radiation production in long implosions of structured gas-puff z pinch loads from large initial radius. Phys Rev Lett 95:105001
46. Atkinson Dean B, Smith Mark A (1995) Design and characterization of pulsed uniform supersonic expansions for chemical applications. Rev Sci Instrum 66(9):4434–4446
47. Braginskii SI (1965) Transport processes in a plasma. 1:205
48. Spitzer L (1956) Physics of fully ionized gases. In: Interscience tracts on physics and astronomy. Interscience Publishers
49. Bennett WH (1934) Magnetically self-focussing streams. Phys Rev 45:890–897
50. Kramida A, Ralchenko YV, Reader J, NIST ASD Team (2015) NIST atomic spectra database (version 5.3)
51. Bethe HA, Salpeter EE (2013) Quantum mechanics of one-and two-electron atoms. In: Dover books on physics. Dover Publications
52. Griem HR (1997) Principles of plasma spectroscopy. In: Cambridge monographs on plasma physics. Cambridge University Press
53. Ralchenko YuV, Maron Y (2001) Accelerated recombination due to resonant deexcitation of metastable states. J Quant Spectr Rad Transfer 71:609
54. Salzmann D (1998) Atomic physics in hot plasmas. In: International series of monographs on physics. Oxford University Press
55. Karzas WJ, Latter R (1961) Electron radiative transitions in a coulomb field. Astrpophysical J Suppl 6:167
56. Tessarin S, Mikitchuk D, Doron R, Stambulchik E, Kroupp E, Maron Y, Hammer DA, Jacobs VL, Seely JF, Oliver BV, Fisher A (2011) Beyond zeeman spectroscopy: magnetic-field diagnostics with stark-dominated line shapes. Phys Plasmas 18(9):093301
57. Stollberg C, Stambulchik E, Duan B, Gigosos MA, González Herrero D, Iglesias CA, Mossé C (2018) Revisiting stark width and shift of He II P$\alpha$. Atoms 6(2):23
58. Golingo RP, Shumlak U (2003) Spatial deconvolution technique to obtain velocity profiles from chord integrated spectra. Rev Sci Instrum 74(4):2332–2337
59. Biswas S, Johnston MD, Doron R, Mikitchuk D, Maron Y, Patel SG, Kiefer ML, Cuneo ME (2018) Shielding of the azimuthal magnetic field by the anode plasma in a relativistic self-magnetic-pinch diode. Phys Plasmas 25(11):113102
60. Doron R, Rubinstein B, Citrin J, Arad R, Maron Y, Fruchtman A, Strauss HR, Mehlhorn TA (2016) Electron density evolution during a fast, non-diffusive propagation of a magnetic field in a multi-ion-species plasma. Phys Plasmas 23(12):122126
61. Rubinstein B, Doron R, Maron Y, Fruchtman A, Mehlhorn TA (2016) The structure of a magnetic-field front propagating non-diffusively in low-resistivity multi-species plasma. Phys Plasmas 23(4):040703
62. Hamdi R, Nessib NB, Sahal-Brechot S, Dimitrijevic MS (2014) Stark widths of Ar III spectral lines in the atmospheres of subdwarf B stars. Adv Space Res 54(7):1223 – 1230. (Spectral line shapes in astrophysics and related phenomena)

63. Dimitrijevic MS (1988) Electron-impact widths of doubly and triply charged ion lines of astrophysical importance. Astron Astrophys Suppl Ser 76:53–59
64. Sahal-Brehot S, Dimitrijevic MS, Moreau N (2015) Stark-B database
65. Konjević N, Lesage A, Fuhr JR, Wiese WL (2002) Experimental stark widths and shifts for spectral lines of neutral and ionized atoms (a critical review of selected data for the period 1989 through 2000). J Phys Chem Ref Data 31(3):819–927
66. Cowan RD (1981) The theory of atomic structure and spectra, vol 3. University of California Press
67. Davara G, Gregorian L, Kroupp E, Maron Y (1998) Spectroscopic determination of the magnetic field distribution in an imploding plasma. Phys Plasmas 5(1):1068–1075

# Chapter 2
# Experimental Setup

The experimental setup is designed to enable the systematic investigation of the axial and azimuthal magnetic fields evolution and plasma properties during Z-pinch implosion in a wide range of plasma and magnetic field parameters.

A quasi-static, nearly uniform axial magnetic field of up to 0.4 T is generated by a pair of Helmholtz coils (see Fig. 2.1 for the schematic description of the components located inside the vacuum chamber). Each coil has a radius of 50 mm and is driven by a circuit with relatively slow rise-time (rise time $\approx$5 ms) to allow for the magnetic field to penetrate into the anode-cathode (A-K) gap through the conducting electrodes. Distance between centers of the coils is equal to the radius of coils in order to achieve uniform distribution of $B_{z0}$ inside the A-K gap.

Subsequently, the gas load is injected into a 10-mm wide anode-cathode gap by a fast gas-puff system with a converging-diverging nozzle. The nozzle forms a hollow cylindrical gas shell with an external diameter of $\approx$38 mm and an internal diameter of $\approx$14 mm (see Fig. 2.2). In all of the experiments presented here, only shell nozzle with argon gas is used. The discharge is initiated when the gas load reaches a selected value between 10–30 $\mu$g/cm. In addition, the gas-puff system has an on-axis jet nozzle that is connected to a separate gas plenum. This feature allows for the introduction of a gas-dopant into the plasma axis region. Furthermore, it is possible to mount a solid target onto the jet nozzle and to use laser ablation for introducing a dopant along the plasma axis. This feature gives further flexibility in selecting suitable species and atomic transitions for the various measurements.

The current is generated by simultaneously discharging four, high voltage, 4-$\mu$F capacitors connected in parallel, driving a 300-kA current pulse with a rise time of 1.6 $\mu$s. The schematic diagram of electrical circuit of the pulse-power generator and plasma is shown in Fig. 2.3. The discharge results in a gas breakdown and the produced plasma carries a current that exerts a $\vec{j} \times \vec{B}$ force in the inward radial direction, compressing the plasma and the axial magnetic field.

© Springer Nature Switzerland AG 2019
D. Mikitchuk, *Investigation of the Compression of Magnetized
Plasma and Magnetic Flux*, Springer Theses,
https://doi.org/10.1007/978-3-030-20855-4_2

**Fig. 2.1** Schematics of components inside the vacuum chamber

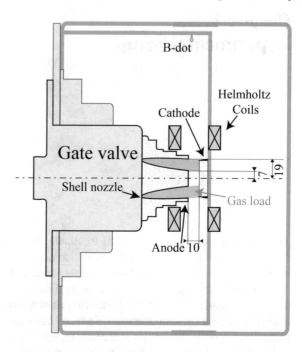

**Fig. 2.2** Nozzle geometry. Part 1: convergent section with subsonic flow. Part 2: neck with Mach 1 flow. Part 3: divergent section with supersonic flow

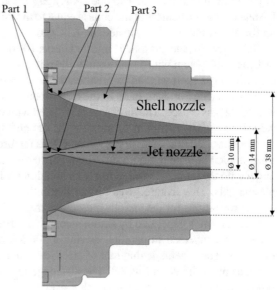

**Fig. 2.3** Electrical circuit diagram of the pulsed-power generator and plasma. $L_{PPS}$ and $R_{PPS}$ represent the total inductance and resistance of the pulsed-power generator, respectively

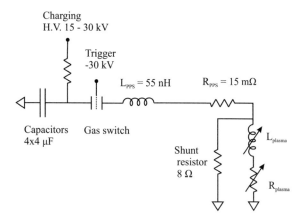

## 2.1 Diagnostic Setups

To study the axial and azimuthal magnetic fields evolution and the plasma parameters, spectroscopic methods were used. Three different spectroscopic setups were employed. Figure 2.4 presents the optical setup used for the simultaneous measurements of the axial and azimuthal magnetic fields at $z = 3.5$ mm (anode is defined as $z = 0$). The simultaneity of the measurements is important in our experiments due to the irreproducibility that characterizes experiment of high-current pulses. This was achieved by employing two UV-visible spectroscopic systems that simultaneously observed line emissions along various chords of the cylindrical plasma. One system includes a 1-m spectrometer equipped with a 2400 grooves/mm grating, which for a typical slit width of 50 μm, provides a spectral resolution of 0.4 Å. The second system includes a 0.5-m imaging spectrometer, equipped with a 1800 grooves/mm grating, which for a typical slit width of 50 μm, provides a spectral resolution of 0.9 Å. For the light detection, the output slit of each of the spectrometers is coupled to a single-gated (3 ns) ICCD camera. The resolution of each spectrometer is mainly defined by the pixel size of the ICCD camera and by the performance of the camera's electrons and light optics.

The $B_z$ measurement is based on the line-width comparison of $\pi$ and $\sigma$ polarization components emitted from dopant transitions. The dopant is generated along the imploding plasma axis by laser ablation ($\lambda = 1064$ nm, $t_{pulse} = 7$ ns, $E_{pulse} = 300$ mJ) of an aluminum target. The $\pi$ and $\sigma$ polarization components were recorded simultaneously on a single detector by exploiting the imaging properties of the 0.5-m spectrometer. The collected light from the dopant plasma is split into $\pi$ and $\sigma$ components by a polarizing beam-splitter and each component is imaged on the upper and lower part of the spectrometer's slit, respectively (see optical system at the right side of Fig. 2.4).

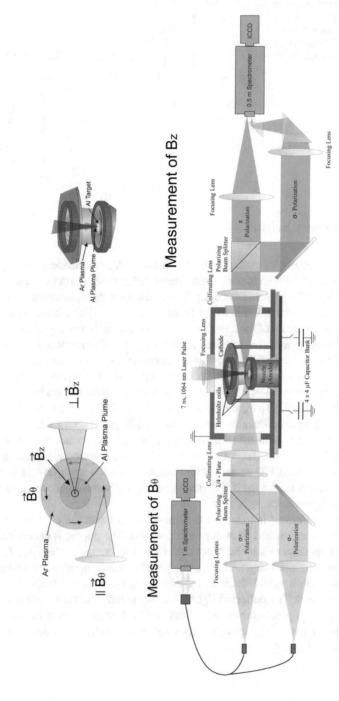

**Fig. 2.4** Spectroscopic setup for simultaneous measurements of the axial and azimuthal magnetic fields

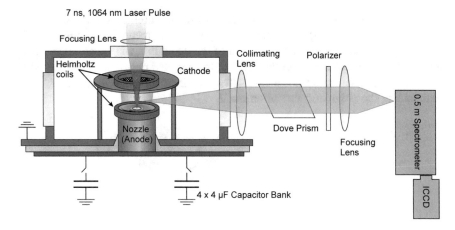

**Fig. 2.5** Spectroscopic setup for measurements of the axial magnetic fields distribution

The $B_\theta$ determination is based on the relative shift between $\sigma^+$ and $\sigma^-$ polarization components of light emitted from the imploding plasma shell. Both polarization components were recorded simultaneously on a single detector by using a bifurcated optical fiber imaged on the 1-m spectrometer (see optical system in the left of Fig. 2.4). While the plasma cross section is imaged on each of the bifurcated ends, composed of 50 fibers, the combined end ensures all 100 fibers are viewed by the spectrometer. The collected light from the imploding plasma is phase shifted by a quarter-wave plate, and subsequently split into $\sigma^+$ and $\sigma^-$ components by a polarizing beam-splitter. Each of the polarization components is imaged on a separate fiber bundle end. We note, the spectral line used for the $B_\theta$ measurement also allows for the simultaneous determination of electron density ($n_e$) from the Stark broadening.

A more detailed description of the spectroscopic methods utilized for the axial and azimuthal magnetic fields measurements is given in Sects. 1.3.2 and 3.1.

Figure 2.5 presents the optical setup used for the measurements of the axial magnetic field distribution on the symmetry axis in $0 < z < 5$ mm range. Different from the setup presented in Fig. 2.4, that was designed to obtain $B_z$ at specific $z$ location, here the measurements are aimed to obtain spectrum for each polarization at all $z$-positions simultaneously. The $\pi$ and $\sigma$ polarizations were measured in different shots in order to collect more light from the dopant emission at $z \sim 5$ mm, where the light intensity was very low. In order to reduce the effect of the shot-to-shot irreproducibility, the measurements were repeated several times and then averaged. As shown in Fig. 2.5, the light from the dopant plasma column, in the selected polarization, was imaged (using two lenses, a polarizer and dove prism) along spatially resolved spectrometer's slit. The spectrometer's output port was coupled to the gated ICCD camera.

Figure 2.6 presents the optical setup used to measure simultaneously the plasma parameters along various chords of the cylindrical plasma. This spectroscopic setup was designed to consist only reflecting optics (i.e. mirrors) in order to exclude

chromatic aberration that significantly distort the measurements when wide spectral region is used, as required here for the plasma parameters determination. The
imploding plasma was imaged on the slit of the 0.5 m imaging spectrometer using
a concave mirror. As shown in the Fig. 2.6, the concave mirror's axis is rotated by
~6°, in order to separate the optical paths of the incident and reflected rays, while
maintaining the astigmatic aberration small.

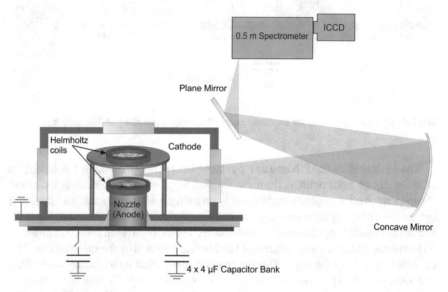

**Fig. 2.6**  Spectroscopic setup for measurements of the plasma parameters

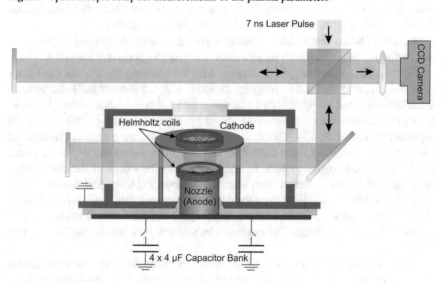

**Fig. 2.7**  Interferometric setup

To study the instabilities evolution for different initial $B_{z0}$, 2D visible imaging and interferometry are used. The plasma is imaged from the radial direction by a 300-mm focal-length lens with $3 \times$ demagnification. The plasma self-emission filtered to the 4000–6000 Å region is recorded by an ICCD camera with a gate time of 5 ns.

The interferometric system, shown in Fig. 2.7, consists of a Michelson interferometer, using the second harmonic (532 nm) of a Q-switched Nd:YAG laser with a pulse duration of 7 ns and a CCD camera.

## 2.2   Initial Conditions Characterization

### 2.2.1   Initial Gas Distribution and Time-Evolution Measurements

Interferometric and Planar Laser Induced Fluorescence (PLIF) methods are employed to examine the gas-valve and nozzle performance, by the determination of the gas distribution and its evolution in the anode-cathode gap.

The interferometric setup utilizes a Mach-Zender interferometer with a 100 mW CW laser at 532 nm. A photodiode is placed at a certain point in the region of interference and records the intensity of the fringes as a function of time. This measurement gives us the temporal evolution of the gas density integrated along the probe-beam path. Figure 2.8 shows a gas evolution measured at the distance of 5 mm from the nozzle outlet and integrated along the probe-beam passing through the diameter of the gas column. The black curve represents the phase shift evolution when the nearly transperent cathode mesh is mounted at the distance of 10 mm from the nozzle (simulating the condition of the experiment). The red curve represents the phase shift evolution without the cathode mesh. In both measurements the shell-nozzle plenum was filled with $CO_2$ gas at 1 atm and the jet-nozzle plenum was empty. These curves show clearly that in both cases the gas evolution is similar for the first $180 \mu s$ (t=0 is set by a pin-trigger discharge). After $180 \mu s$, part of the gas that is back-scattered from the cathode starts to effect the gas distribution along the probe-beam. Pin-trigger is an electrode located in the shell-nozzle convergent section (see Fig. 2.2) and prior to the Z-pinch discharge is set to 2 kV relatively to the nozzle. The breakdown between the pin-trigger and the nozzle occurs when the gas starts flowing into the nozzle inlet. In the experiment, this pulse is used for triggering delay generator that generates the time-sequence of Z-pinch discharge and instruments operation. Such triggering sequence guarantees reproducibility of initial gas distribution when the Z-pinch discharge is initiated.

Figure 2.9 presents a comparison between the phase shift evolution obtained for shell- and jet-puff (black curve), with one obtained for shell-puff only (red curve). In both cases the measurements are performed without the cathode mesh. The jet-puff only is not presented, because the pin-trigger electrode can be located only along the shell-nozzle gas flow. From Fig. 2.9 one can see that the gas from the jet-nozzle

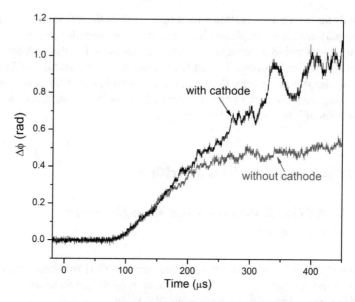

**Fig. 2.8** Interferometric measurements of the injected shell-gas evolution. Both curves represent the probe beam phase shift evolution measured at 5 mm distance from the nozzle. The black curve is the phase shift with the cathode mesh mounted at 10 mm distance from the nozzle and the red curve is the phase shift evolution without the cathode

appears in the anode-cathode gap about 20 $\mu$s earlier than the shell-nozzle gas, and also reaches steady state earlier.

In the PLIF measurement a *planar* laser beam is directed through the gas flow, doped with a molecular tracer. The wavelength of the incident beam is tuned to excite a particular transition of the molecular tracer. Subsequently, the tracer fluorescent emission is recorded by the ICCD camera placed perpendicularly to the beam. In the present work, acetone tracer is used and irradiated by $\lambda = 354.7$ nm laser beam (3rd harmonic of Q-switched Nd:YAG laser). PLIF measurement gives the spatial gas distribution (assuming that the tracer is uniformly distributed inside the gas, therefore, the light-intensity is proportional to the gas density) at a certain time (the time scale of the fluorescence emission (few ns) is much smaller than the time scale of the gas flow (few tens $\mu$s)). Figure 2.10 presents a PLIF image recorded by ICCD camera at 180 $\mu$s after the pin-trigger signal. It is seen from the image that at t=180 $\mu$s the jet- and shell-puff inside the anode-cathode gap (up to 6 mm) nearly preserve the form and dimensions set by the nozzles' outlet geometry. At the region $z > 6$ mm the effects of back-scatterd gas and radial gas expansion become notable.

Due to the planar geometry of the exciting laser beam, the PLIF measurements give information on the gas-distribution as a function of radius (at certain time) without the need of applying inverse Abel-transform procedure. Therefore, combining the PLIF results with the cordal-integrated interferometric measurements (Fig. 2.9), one obtains the mass-per-length time-evolution of the shell- and jet-puffs. This parameter

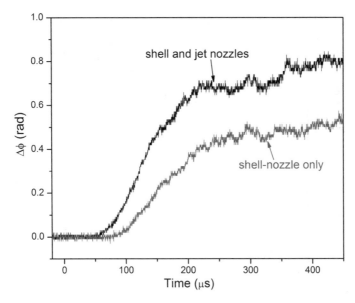

**Fig. 2.9** Interferometric measurements of the injected gas evolution. Both curves represent the probe-beam phase-shift evolution measured at 5 mm distance from the nozzle. Black curve is the phase shift as a function of time when both nozzle plena were filled with $CO_2$ gas at 1 atm and the red curve is the same but only shell-nozzle plenum was filled with $CO_2$ gas at 1 atm

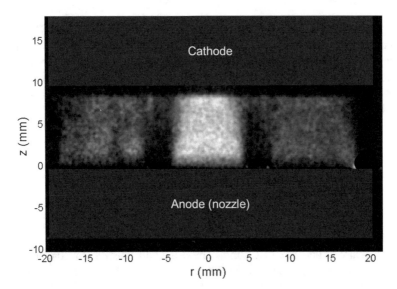

**Fig. 2.10** ICCD PLIF image recorded at 180 $\mu$s after the pin-trigger signal. Gas pressure ($CO_2$) in both plena was set to 2 atm

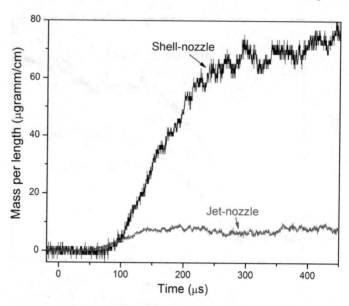

**Fig. 2.11** $CO_2$ gas mass-per-length evolution at a distance of 5 mm from the nozzle outlet. The black curve represents the mass per length time dependence of the shell-puff. The red curve represents the mass per length time dependence of the jet-puff

is important for the plasma and magnetic field dynamic simulations. Figure 2.11 presents the $CO_2$ gas mass per length evolution at the distance of 5 mm from the nozzle outlet. The initial pressure in each of the plena was set to 1 atm and the interferometric measurements are performed without the cathode mesh (using the results of Fig. 2.8 one can assume that the mass-per-length evolution with cathode mesh is the same as without the mesh for the first 200 ns).

## 2.2.2  B-Dot Calibration for Discharge Current Measurements

The measurement of the discharge current in the circuit is performed by a B-dot coil placed between the A-K gap at $z = 5$ mm ($z = 0$ is at the outlet of the nozzle), and the return-current path at $r \approx 120$ mm. The B-dot coil consists of a conductive loop connected to a voltage recording device. The voltage on the coil is induced due to changes in the magnetic flux through it. Magnetic flux passing through the loop is proportional to the current flowing in the load, therefore using the initial condition $I(t = 0) = 0$ A and the time integral of the voltage measurements on the B-dot we obtain a curve proportional to the current evolution. The absolute calibration of the B-dot is performed using the fit to the measured time-integrated B-dot signal as explained in the next paragraph. For the calibration purposes the anode-cathode gap was bridged by an aluminum conductor that simulates the plasma column electric conductivity.

The pulse-power system shorted at A-K gap is a RLC circuit (see Fig. 2.3), therefore its current can be simulated by a damped sinus function: $I(t) = Ae^{-t/\tau}\sin(\omega t)$, where $\tau = \frac{2L}{R}$; $\omega = \sqrt{\frac{1}{LC} - \frac{R^2}{4L^2}}$ ($R$, $L$, and $C$ are the circuit resistance, inductance, and capacitance, respectively) . From the best fit to the integrated B-dot signal one finds the damping constant $\tau$ and frequency $\omega$, and together with the measured capacity $C$ and the charging voltage of the capacitors $V_0$ it is possible to obtain the circuit $R$, $L$, current amplitude, and rise time. The total capacitance of the current-generating circuit is $C = 16\,\mu F$, and the capacitors are charged to $V_0 = 20$ kV. The best fit shown in Fig. 2.12 is obtained for: $\tau = 10^{-5}$ s; $\omega = 1.043 \times 10^6$ s$^{-1}$. Therefore the circuit parameters are: $R = 11.5$ m$\Omega$, $L = 60$ nH, $t_{rise} = 1.5\,\mu s$, $I_{max} = 280$ kA. The proportionality constant $\alpha$ (or the calibration constant between the B-dot and the current generating circuit) is given by: $\alpha = I_{max}/$(maximum of time integrated Bdot signal) $= 4.2 \times 10^{10}$.

Figures 2.13 and 2.14 present the typical traces of the B-dot signal (black curve) of shots with plasma and the calculated (using proportionality constant $\alpha$) discharge currents (red curve). Figure 2.13 shows the discharge with the capacitors charged to 23 kV and without the application of initial axial magnetic field. We note, that the deep in the B-dot trace observed at $t \sim 700$ ns is due to the fast increase of the plasma inductance and in some cases also the plasma resistance during the final stage of the implosion. This causes a significant rise of the plasma impedance and subsequently drop of the current, due to the term $\frac{dL}{dt}I$ in RLC equation with varying inductance: $V = L\frac{dI}{dt} + \frac{dL}{dt}I + RI$. Figure 2.14 shows the discharge with the capacitors charged to 23 kV for $B_{z0} = 0.4$ T. Here, the deep in the B-dot trace is not observed due the significant current loss to the low-density peripheral plasma when $B_{z0}$ is applied,

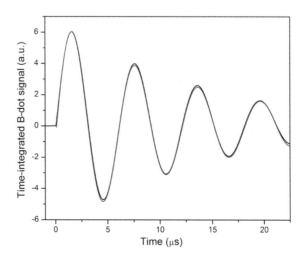

**Fig. 2.12** Time integrated Bdot signal (black curve), and its simulation (red curve)

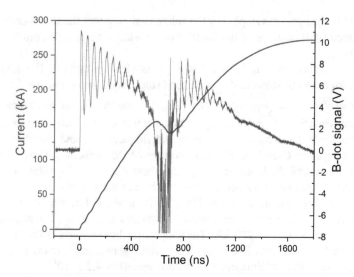

**Fig. 2.13** B-dot trace (black) and calculated current (red) for a typical shot with plasma of the pulsed-power system. The discharge was performed with the capacitors charged to 23 kV and without the application of initial axial magnetic field

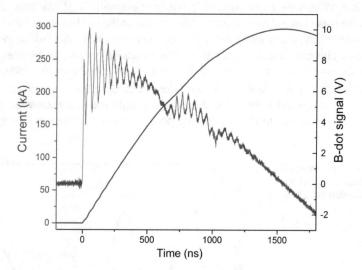

**Fig. 2.14** B-dot trace (black) and calculated current (red) for a typical shot with plasma of the pulsed-power system. The discharge was performed with the capacitors charged to 23 kV and with $B_{z0} = 0.4$ T

as described in the Chaps. 3 and 4. The loss of the current to the peripheral plasma keeps the impedance of the imploding plasma low throughout the implosion. The fast oscillations (period ~50 ns) observed in the B-dot signal are found to be due to the parasitic capacitance, parallel to the plasma, of the pulsed-power-generator feeds (such oscillations of B-dot signals are typical in the pulsed-power devices).

### 2.2.3 Initial Axial Magnetic-Field Evolution Measurements

The external axial magnetic field, $B_{z,ext}$, is generated by a current pulse (rise-time $\approx$4.5 ms) driven through the Helmholtz coils. Since, $B_{z,ext}$ is varying in time, also here a B-dot probe is used for the determination of its evolution inside the A-K gap. A B-dot probe is made of three 7 mm diameter loops, which is mounted inside the A-K gap at $z \approx 5$ mm. The voltage generated in the B-dot is proportional to the rate of the magnetic flux change. The time integration of the B-dot signal and the known encircled area give the axial magnetic field evolution.

$$B_{z,ext}(t) = \frac{1}{NS} \int_0^t V(t')dt' \qquad (2.1)$$

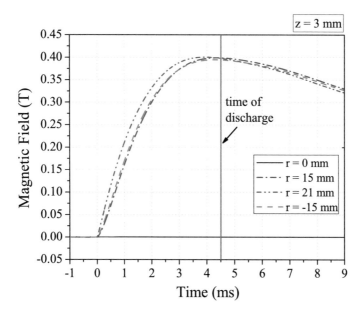

**Fig. 2.15** Evolution of the external axial magnetic field, generated by the Helmholtz coils at different radial positions

$N$ is the number of the loops, $S$ is the area of each loop, and $V(t')$ is the measured B-dot voltage signal.

Figure 2.15 presents the evolution of the external axial magnetic field generated by the Helmholtz coils at $z \approx 3$ mm and different radial positions. The rise time of the axial $B$-field is 4.5 ms, which is much longer than the time scale of the plasma compression ($\sim 1\,\mu$s). Therefore, we consider $B_{z,ext}$ constant during the plasma implosion. The plasma discharge is initiated at the maximum of $B_{z,ext}$, i.e. $B_{z0} = B_{z,ext}(t = 4.5\text{ms})$ It is seen from Fig. 2.15 that at this time $B_{z,ext}$ is most uniform for $r \leq 21$ mm. In the used configuration of Helmholtz coils, the spatial variation of the magnetic field is <2% in the A-K gap ($0 \leq z \leq 10$ mm, $r \leq 21$ mm).

# Chapter 3
# Results

## 3.1 Simultaneous Measurements of the Azimuthal and Axial Magnetic Fields

To understand the evolution of the plasma implosion and the nature of the plasma-$B$-field interaction it is essential to measure the azimuthal and axial magnetic fields evolution during the implosion. The main challenge in the determination of the azimuthal magnetic field in typical imploding plasma experiments comes from the large Stark broadening of the emission lines due to the high electron density that results in the smearing out of the Zeeman-split pattern. To resolve this problem we employed spectroscopic polarization technique based on the measurement of the relative wavelength shift between the $\sigma^+$ and $\sigma^-$ polarizations of the Zeeman components (see Sect. 1.3.2). This approach is applicable when the line-of-sight is parallel to the measured magnetic field. Therefore, here it can be used for $B_\theta$ determination at the outer edge of the plasma shell, radially observed, where the $B_\theta$ is directed along the line-of-sight (see upper left inset in Fig. 2.4). The high sensitivity of this method is due to the fact that both polarization components are measured simultaneously, using a single spectrometer and a single detector, allowing for a reliable and accurate determination of very small wavelength difference between the polarization components. We note that this method is applicable even when $B_z$ is also present, as long as $B_\theta > B_z$, a condition that is valid in our case for the plasma periphery.

The determination of $B_z$ inside the plasma shell poses two major difficulties. The first is the difficulty to distinguish between the axial and the azimuthal magnetic fields along the line of sight, and the second is the absence of light emission from the nearly hollow axial central region of the plasma column. We succeeded to circumvent these problems by introducing a dopant into the central region of the plasma column, employing laser ablation of an aluminum target placed onto the jet-nozzle. Thus, the axial magnetic field evolution and distribution was measured using selected transitions of ions generated from the dopant, while the azimuthal field was measured

© Springer Nature Switzerland AG 2019
D. Mikitchuk, *Investigation of the Compression of Magnetized Plasma and Magnetic Flux*, Springer Theses,
https://doi.org/10.1007/978-3-030-20855-4_3

using transitions in the plasma shell-gas ions. The dopant density must be sufficiently high ($n_e \sim 3 \times 10^{17}$ cm$^{-3}$), so its line emission is clearly observed above the ambient argon plasma continuum. The difficulty in measuring magnetic fields (in the relevant range of 1–5 T) in such densities is the smearing out of the Zeeman splitting by the Stark broadening. To overcome this difficulty, the magnetic field is determined using the polarization technique based on the measurements of the relative width difference between the $\pi$ and $\sigma$ polarizations of the Zeeman components [1] (see Sect. 1.3.2). The dopant technique allows for utilizing a large variety of materials as dopant and selecting the most suitable atomic transitions for the $B_z$ measurements in the studied range of parameters. Here, the Al III 4s $^2S_{1/2}$–4p $^2P_{1/2}$ transition at $\lambda = 5722.7$ Å was selected for the axial B-field determination due to its relatively high emission intensity, high sensitivity to the magnetic field, and since it is not blended with the argon lines. Also, this atomic transition doesn't involve the ground state and thus opacity effects are negligible.

Figure 3.1 presents an example of a spectroscopic image recorded at $t = 805$ ns (where the beginning of the current is defined as $t = 0$) for $B_{z0} = 0.4$ T and at $z = 3.5$ mm away from the nozzle. The upper and bottom parts of Fig. 3.1 present the $\pi$- and the $\sigma$- components of the aluminum emission, respectively. It is seen in Fig. 3.1 that the Al dopant resides in the $-0.6 < y < 0.6$ mm region therefore providing an information on the $B_z$-field, averaged over $\Delta y \approx 1$ mm. The Al lines seen on the image are Al III 4s $^2S_{1/2}$–4p $^2P_{3/2}$ at $\lambda = 5696.6$ Å and Al III 4s $^2S_{1/2}$–4p $^2P_{1/2}$ at $\lambda = 5722.7$ Å. Since the stronger Al III transition at $\lambda = 5696.6$ Å is slightly blended with weak argon lines, only the transition at $\lambda = 5722.7$ Å is used for the $B_z$ determination. The extraction of $B_z$ from the measured data involves three main steps:

1. Lineouts generation of the Al III 4s $^2S_{1/2}$–4p $^2P_{1/2}$ transition: In order to improve the signal to noise ratio, the lineouts are integrated across the dopant plume. The axial magnetic field is assumed to be nearly homogeneous in this region (dopant extent in radial direction is $\sim 1$ mm).
2. $B_z$ determination: The magnetic field is extracted by simultaneous fitting of the measured $\pi$ and $\sigma$-polarization lineshapes, with the convolution of the calculated Zeeman pattern, Lorentzian (due to Stark broadening), and Gaussian (due to instrumental and Doppler broadenings) profiles. The parameters of the simulations are kept the same for both $\pi$ and $\sigma$. It is emphasized that only the different contributions of the Zeeman pattern to the $\pi$ and $\sigma$ polarizations causes a difference in the simulated lineshapes and allows for the $B_z$-field determination. The Gaussian contribution to the lineshapes is limited to $0.6 < FWHM_{Gauss} < 0.8$ Å, the lower limit results from the known instrumental broadening, and the upper limit from the lineshape measurements of the freely expanding dopant plasma.
3. Error estimation: The lower bound of the $B_z$ error bar is determined in two steps. First, the lineshape of the measured $\pi$-polarization component, that has a smaller Zeeman splitting, is fitted by Voigt distribution with $B_z = 0$. In this way we obtain the largest possible contribution of the Gaussian and Lorentzian distributions to the lineshape. In the second step, we find the best fit for the lineshape of the

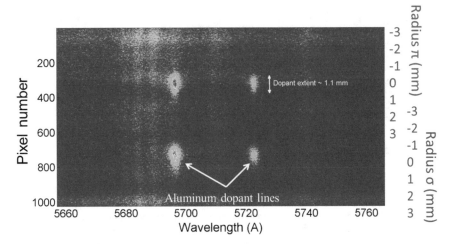

**Fig. 3.1**  Spectral image of the Al III $4s$–$4p$ doublet at $\lambda = 5696.6$ Å and $\lambda = 5722.7$ Å, recorded at $t = 805$ ns, when the plasma radius is 6 mm. The upper and bottom parts of the image consist of the $\pi$ and $\sigma$ polarized components of the dopant spectrum, respectively

$\sigma$-polarization component using the upper bound for the Gaussian and Lorentzian parameters found in step one, and $B_z$ is a free parameter. This procedure gives the lowest possible $B_z$ for each data point.

The determination of the upper bound of the $B_z$ error bars is similar. First, we find the smallest possible contribution of the Gaussian and Lorentzian distributions to the lineshape. The $FWHM$ of the Gaussian was set to be the instrumental broadening of 0.6 Å, and the $FWHM$ of the Lorentzian was set to be of the Stark broadening of the Al dopant plasma in free expansion ($B_z = 0$) that was 1.3 Å (it was seen that the dopant plasma density is lowest for dopant expansion without $B_z$-field). The upper bound of $B_z$ is then determined by simultaneous fitting of the measured $\pi$ and $\sigma$-polarization lineshapes with fixed $FWHM_{Gaussian} = 0.6$ Å, $FWHM_{Lorentz} = 1.3$ Å, and $B_z$ as a free parameter.

An example of the lineshape analysis performed for the $\pi$ and $\sigma$ components of the Al III $4s$ $^2S_{1/2}$–$4p$ $^2P_{1/2}$ transition, obtained from the spectral image shown in Fig. 3.1, is presented in Fig. 3.2. The errors in the lineshape data points are determined from the standard deviation of the binned (performed along the $y$-direction) pixel signals. The clear difference in the widths of two polarization components reveals the existence of an axial magnetic field in the measured region. The simultaneous best fit simulation for $\pi$ and $\sigma$ is obtained for $B_z = 2.9$ T, $FWHM_{Lorentz} = 1.7$ Å, and $FWHM_{Gauss} = 0.6$ Å.

As described in Sects. 1.3.2 and 1.2, $B_\theta$ was measured simultaneously with $B_z$ using the relative shift between the $\sigma^+$ and $\sigma^-$ polarization components of the Zeeman pattern of the Ar III $(^4S)4s$ $^5S_2$–$(^4S)4p$ $^5P_2$ transition. Figures 3.3 and 3.4 are examples of two spectroscopic images of the argon plasma spectrum recorded at $t = 595$

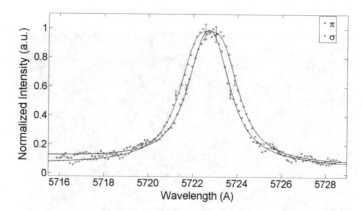

**Fig. 3.2** Comparison between the $\pi$ polarization (blue circles) and $\sigma$ polarization (red circles) lineshapes recorded at $t = 805$ ns for $B_{z0} = 0.4$ T, together with their best fit simulations (solid lines) obtained for $B_z = 2.9$ T, $FWHM_{Lorentz} = 1.7$ Å, and $FWHM_{Gauss} = 0.6$ Å

ns for $B_{z0} = 0$ and $t = 815$ ns for $B_{z0} = 0.4$ T, respectively. Both images show the cord-integrated emissions from the plasma cross section at $z = 3.5$ mm. The lines seen in these figures originate from Ar III $(^4S)4s\,^5S_2$–$(^4S)4p\,^5P_2$ at $\lambda = 3301.9$ Å and $(^4S)4s\,^5S_2$–$(^4S)4p\,^5P_1$ at $\lambda = 3311.2$ Å transitions, together with several unidentified lines that are likely belong to Ar IV, considering their time evolution. For the $B_\theta$ determination, the Ar III transition at $\lambda = 3311.2$ Å was selected due to its relatively high intensity, high sensitivity to the magnetic field, and the fact that the atomic transition doesn't involve the ground state and thus opacity effects are negligible. Only lineouts generated at the periphery of the plasma were used for the $B_\theta$ extraction, in order to ensure that the line of sight is parallel to $B_\theta$-field and to obtain the $B$-field due to the total current flowing through the imploding argon plasma. The data analysis process is similar to the one described earlier for the $B_z$ determination. First, the lineout of the Ar III transition at $\lambda = 3301.9$ Å was generated by integration over the radii in which the emission changes from $\sim5$ to $\sim30\%$ of the peak emission, after inverse Abel transform. This integration corresponds to $\Delta r \lesssim 0.5$ mm, whereas the plasma shell width is about 3 mm. Then, the $\sigma^+$ and $\sigma^-$-polarization components are fitted simultaneously taking into account the convolution of the calculated Zeeman pattern, Lorentzian, and Gaussian profiles. $B_\theta$ is then determined from the relative wavelength shift between the two polarization components.

An example of the lineshape analysis performed for the $\sigma^+$ and $\sigma^-$ components of the Ar III $(^4S)4s^5S_2$–$(^4S)4p^5P_2$ transition, obtained from the spectral images shown in Figs. 3.3 and 3.4, are presented in Figs. 3.5 and 3.6, respectively. As was already mentioned, the errors in the lineshape data points are determined from the standard deviation of the binned (performed along the $y$-direction) pixel signals. In both figures, we clearly see the relative wavelength shift between the two spectral features due to the presence of the magnetic field. The best fit for the lineshapes and the wavelength shift presented in Fig. 3.5 is obtained for $B_\theta = 4.1$ T, $FWHM_{Lorentz} = 0.55$ Å

**Fig. 3.3** Spectral image of the Ar III $(^4S)4s\,^5S_2$–$(^4S)4p\,^5P_2$ transition at $\lambda = 3301.9$ Å recorded at $t = 595$ ns for $B_{z0} = 0$. The upper and bottom parts of the image consist of the $\sigma^+$ and $\sigma^-$ polarized components of the argon plasma spectrum, respectively. The dashed line represents the unshifted (zero $B_\theta$-field) position of the used line

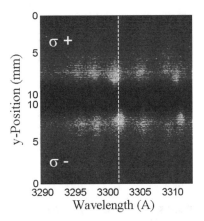

**Fig. 3.4** Spectral image of the Ar III $(^4S)4s\,^5S_2$–$(^4S)4p\,^5P_2$ transition at $\lambda = 3301.9$ Å recorded at $t = 815$ ns for $B_{z0} = 0.4$ T. The upper and bottom parts of the image consist of the $\sigma^+$ and $\sigma^-$ polarized components of the argon plasma spectrum, respectively. The dashed line represents the unshifted (zero $B_\theta$-field) position of the used line

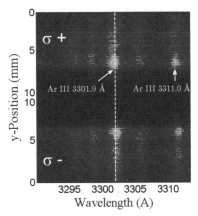

that corresponds to $n_e \sim 4.6 \times 10^{17}$ cm$^{-3}$, and $FWHM_{Gauss} = 0.5$ Å that is mostly due to the instrumental broadening ($\sim$0.4 Å). The best fit for the lineshapes and the wavelength shift presented in Fig. 3.6 is obtained for $B_\theta = 1.9$ T, $FWHM_{Lorentz} = 1$ Å that corresponds to $n_e \sim 7 \times 10^{17}$ cm$^{-3}$, and $FWHM_{Gauss} = 0.5$ Å.

Figure 3.7 presents the measurements of $B_\theta$ as a function of the plasma radius, together with the calculated $B_\theta$ using the measured total current and the plasma radius from which the Ar III lineout is generated, as explained above. Since for the magnetic fields in the present study the Zeemann splitting of the lines used for the $B$-field diagnostics is small compared to the spin-orbit interaction energies of the multiplet, the distance between the line centers of $\sigma^+$ and $\sigma^-$ components is proportional to $B_\theta$ magnitude. Therefore, the error bars of the measured $B_\theta$ were determined from the confidence range of the lineshape centers of the best-fit to the $\sigma^+$ and $\sigma^-$ polarization components.

All the measurements presented in Fig. 3.7 were performed for $B_{z0} = 0$, at $z = 3.5$ mm from the anode (nozzle). Due to the fast ionization processes in implosion without axial magnetic field, the $B_\theta$ measurements based on the Ar III transition are limited to

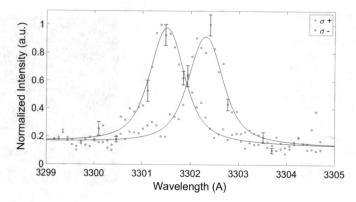

**Fig. 3.5** The $\sigma^+$ polarization (blue circles) and $\sigma^-$ polarization (red circles) lineshapes recorded at $t = 595$ ns in implosion without initial axial $B$-field, together with their best fit simulations (solid lines) obtained for $B_\theta = 4.1$ T, $FWHM_{Lorentz} = 0.55$ Å, and $FWHM_{Gauss} = 0.5$ Å

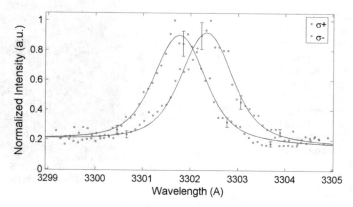

**Fig. 3.6** The $\sigma^+$ polarization (blue circles) and $\sigma^-$ polarization (red circles) lineshapes recorded at $t = 815$ ns for $B_{z0} = 0.4$ T, together with their best fit simulations (solid lines) obtained for $B_\theta = 2$ T, $FWHM_{Lorentz} = 1$ Å, and $FWHM_{Gauss} = 0.5$ Å

$500 < t < 600$ ns, corresponding to radii range $8.5 < R < 10.5$ mm. In the future, we plan to employ Ar IV transitions in order to extend the $B_\theta$ determination to later times. $B_\theta$ from the current is calculated by

$$B_\theta = \frac{\mu_0 I(t)}{2\pi R(t)}, \qquad (3.1)$$

where $I(t)$ and $R(t)$- are the total current and the plasma radius at time $t$, respectively. Here, the calculated $B_\theta$ is the expected azimuthal magnetic field assuming all the current flows within the plasma radius. It is seen in the figure that for implosions without initial $B_z$ the measured and calculated data points agree well. This shows that all the current is flowing within the imploding argon shell. This result agrees

**Fig. 3.7** Measured $B_\theta$ (red dots) as a function of the argon plasma radius measured at $z = 3.5$ mm from the anode for $B_{z0} = 0$, together with the calculated $B_\theta$ (blue triangles) using the total current and plasma radius. The upper scale shows the typical times that correspond to each plasma radius

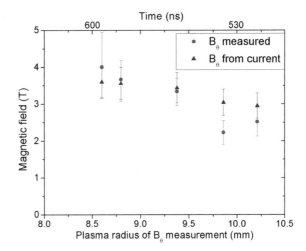

with a previous study [1] carried out in Z-pinch implosion (without axial $B$-field) showing that the entire current flows within the imploding plasma.

Figure 3.8 presents the simultaneous measurements of $B_z$ and $B_\theta$ as a function of the plasma radius, together with the calculated $B_\theta$ using the measured total current and plasma radius (assuming all the current flows in the observed plasma radius). All the measurements presented in Fig. 3.8 were performed for $B_{z0} = 0.4$ T, at $z = 3.5$ mm from the anode (nozzle). The plasma radius is determined from the spectral images of the Ar III line and is defined to be the radius at which $B_\theta$ is measured at the periphery of the Ar III line emission, as described above. The calculation of $B_\theta$ from the current is given by Eq. 3.1.

Figure 3.8 presents the current flowing within the outer radius of the imploding plasma (red dots) as a function of the plasma radius, together with the measured total current (blue triangles). The current within the imploding plasma is calculated using Eq. 3.1 with measured $B_\theta$ at the outer radius of the argon plasma.

We now describe several important results observed in Figs. 3.8 and 3.9. The first refers to the $B_\theta$ measurement. It is seen from the figure that the measured $B_\theta$ is much smaller ($\sim$4 times) than the expected value based on the total current and the observed plasma radius. This result is in contrast to the measurements performed without initial $B_z$ that showed a good agreement between the measured and calculated $B_\theta$ values. The unexpectedly small magnitude of $B_\theta$ observed at the plasma outer radius for discharges with $B_{z0} = 0.4$ T shows that most of the current doesn't flow within the imploding argon plasma and even decreases with time as can be seen in Fig. 3.9. It means that at times close to the stagnation ($t > 750$ ns), $\sim$75% of the total current flows outside, at radii higher than the imploding plasma radius, raising a question regarding the current distribution at these high radii. To resolve this puzzle, a new spectroscopic setup to measure the azimuthal magnetic field at radii larger than the imploding plasma radius was recently built. Recent results revealed the existence of a low-density peripheral plasma and a significant current flow in this region (however

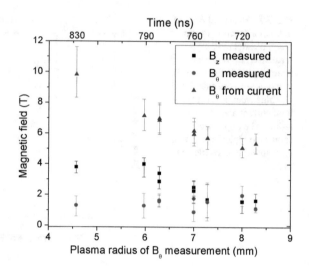

**Fig. 3.8** Measured $B_\theta$ (red dots) and $B_z$ (black squares) as a function of the argon plasma radius measured at $z = 3.5$ mm from the anode for $B_{z0} = 0.4$ T, together with the calculated $B_\theta$ (blue triangles) using total current and plasma radius. The upper scale shows the typical times that correspond to each plasma radius

these measurements are outside the scope of the present research and can be found in [2]). This is a very important finding because it shows that the presence of $B_z$ dramatically affects the current distribution. The discussion on this result will be given in Chap. 4.

Another important result shown in Fig. 3.8 is that at times close to stagnation ($t_{stagnation} \sim 815$ ns) $B_z$ is almost $3\times$ higher than $B_\theta$. The observation that $B_z > B_\theta$ at stagnation is in agreement with theoretical calculations presented in Sect. 1.2.1, where always $B_{z,stagnation} > B_{\theta,stagnation}$ due to the plasma inertia effect. However, in contrast to the theoretical calculations the measured maximum $B_z$ is much lower than the one predicted by the calculations (see Fig. 1.7). This relatively low compression factor observed in the experiment is due to relatively small compressing $B_\theta$, as described above.

The third important result of these measurements is the significant delay of the stagnation time for implosion with $B_{z0} = 0.4$ T in comparison to implosion without $B_z$. This is in contradiction with the predictions presented in Fig. 1.4. For example, the measured stagnation times of plasma implosion with $B_{z0} = 0$ and $B_{z0} = 0.4$ T are $\sim$665 and $\sim$815 ns, respectively, i.e the implosion with $B_{z0} = 0.4$ T is $\sim$20% longer than the implosion without $B_z$. On the other hand, for these parameters the calculations predict only $\sim$3% difference in stagnation times (see Fig. 1.4 and Table 1.4). The large delay in stagnation time for $B_{z0} = 0.4$ T is due to the observed "current escape" from the imploding plasma to larger radii, resulting in much lower Lorentz force $\mathbf{j} \times \mathbf{B}$ and subsequently smaller acceleration that leads to longer implosion times.

Figure 3.10 summarizes the $B_z$ measurements as a function of the plasma radius, together with the calculated axial magnetic field, assuming $B_z$ is compressed by ideal plasma (zero resistivity) implosion. Here, the plasma radius is assumed to be at the radius of maximum Ar III line emission, obtained by Abel transform inversion of

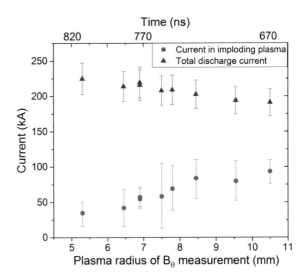

**Fig. 3.9** Current flowing within the imploding plasma (red dots) as a function of the argon plasma radius measured at $z = 3.5$ mm from the anode for $B_{z0} = 0.4$ T, together with the measured total current (blue triangles). The upper scale shows the typical times that correspond to each plasma radius

the Ar III spectral image. Note that the plasma radius definition here is different than in Fig. 3.8 where it is taken at the outer periphery of the plasma). This definition of the plasma radius by the maximum of the Ar III line emission is more relevant for the estimation of the $B_z$ confinement efficiency since the $B_z$-flux diffuses from inside out. The $B_z$ evolution in an ideal plasma implosion is calculated by $B_{z,ideal}(t) = B_{z0} \left(\frac{R_0}{R(t)}\right)^2$, where $R_0$ is the plasma radius at $t = 0$ and $R(t)$ is the plasma radius at time $t$. This formula represents the magnetic-field flux conservation within the $B_z$-confinement radius as shown in Fig. 3.10. We note that the experimental data for $B_z$ consist also points obtained in shots where $B_\theta$ was not measured.

From Fig. 3.10, we see that at maximum compression (plasma radius ~3.4 mm), the measured $B_z$ is ~2 times smaller than the calculated $B_z$ assuming 100% flux conservation. This shows that for the plasma parameters of the present experiment the $B_z$-confinement efficiency is ~50%. The uncertainty of the confinement efficiency is obtained by assuming the largest and smallest possible values of the confinement radius resulting in confinement efficiencies between 25 and 80%. The relatively large uncertainty in the confinement radius determination is due to the width of the plasma shell (FWHM ~2 mm) that is comparable to the plasma radius at stagnation (~3.5 mm).

## 3.2 Axial Magnetic Field Distribution Along $z$-axis

To study the $B_z$ distribution as a function of $z$ and time, the optical arrangement presented in Fig. 2.5 was employed. Also here we used the spectroscopic technique based on the lineshapes width comparison between the $\pi$ and $\sigma$ polarizations of

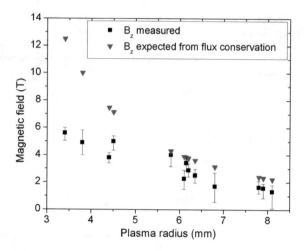

**Fig. 3.10** Measured $B_z$ (black squares) as a function of the argon plasma radius measured at $z = 3.5$ mm from the anode for $B_{z0} = 0.4$ T, together with the calculated $B_z$ (pink triangles) assuming an ideal (zero-resistivity) plasma implosion

the Zeeman splitting. It is important to note that different from the measurements presented in Sect. 3.1, each polarization was recorded in separate shots. Therefore, in order to improve the signal-to-noise ratio and to smooth the effect of shot-to-shot irreproducibility, each analyzed spectrum is obtained by averaging several spectral images recorded at the same time.

An example of two spectral images recorded at $t = 820$ ns with $B_{z0} = 0.4$ T is presented in Fig. 3.11, the left and right panels present the $\pi$ and $\sigma$-polarizations, respectively. The lineouts generated from the spectral images are spatially integrated over a certain $z$ range, determined by the signal to noise ratio in each measurement. For example, in the images recorded at $t > 700$ ns, the spatial integration for the lineouts generated in the $0 < z < 3$ mm region is 0.5 mm, while at $z \sim 5$ mm the integration is performed over $\sim 2$ mm. The lineshape analysis and the error determination is the same as described in Sect. 3.1.

Figure 3.12 shows the measured $B_z$ magnitude on the $z$-axis as a function of $z$ at different times of the plasma evolution. These measurements have been performed during plasma implosion and into stagnation, that occurs at $t_{stagnation} \sim 830$ ns.

An interesting observation seen in Fig. 3.12 is that $B_z$ develops a strong gradient along the $z$-axis for $t \geqslant 770$ ns. The axial magnetic field near the anode is $\sim 2$ times smaller than at the middle of the plasma ($z \sim 5$ mm). This result is consistent with the plasma column shape observed in 2D imaging (6000–7500 Å) as shown in Fig. 3.13 for $t = 825$ ns. It is seen in Fig. 3.13 that at $z = 5$ mm the plasma is compressed to a $\sim 40\%$ smaller radius than in the near anode region. This difference in plasma radii corresponds to $\sim 2.5 \times$ stronger $B_z$ compression at $z = 5$ mm, assuming ideal plasma implosion and taking an initial plasma radius of 19 mm. These observations are explained by the fact that the axial magnetic field lines are frozen in the metal

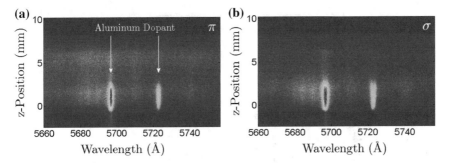

**Fig. 3.11** Spectroscopic images of the aluminum dopant emission recorded at $t = 820$ ns with $\pi$-Zeeman component (**a**) and with $\sigma$-Zemann component (**b**). The measurements are performed with the spatial axis of the image along the plasma symmetry axis

electrodes with $B_z = B_{z0}$. The field inside the metal can not be compressed. Therefore, the spatial shape of the plasma near the anode surface is a result of the transition from the uncompressed $B_z = B_{z0}$ in the electrode to the compressed axial magnetic field further away from the anode surface, see Fig. 3.14. Such transition region is not expected near the cathode made of 1 mm diameter stainless steel wire mesh, since the diffusion length of $B$-field for the time-scale of the implosion is comparable to the diameter of the wire. However, this behavior can not be observed due to a 'knife edge' mounted on the cathode surface that blocks the view of the plasma near the cathode.

Another notable result seen in Fig. 3.12 (most clearly at $z = 4.5$ mm) is that $B_z$ decreases during stagnation (i.e. $B_z(t = 825 \text{ ns}) > B_z(t = 875 \text{ ns})$). The decrease of $B_z$ indicates a diffusion of the axial magnetic field through the confining plasma. The decreasing $B_z$ on axis and the continuous rise in current (that should give rise to larger $B_\theta$) should have led to further compression of the plasma shell. However, the plasma radius (inferred from the 2D and interferometric measurements) remains practically the same. This result indicates a "current escape" not only during the implosion stage, but also from the stagnating plasma (where there is no experimental data for $B_\theta$). We see that further experimental data is required to understand the pressure balance at stagnation.

## 3.3 Effects of Axial Magnetic Field on the Plasma Implosion

The success of a Z-pinch application as an efficient x-ray source or for inertial confinement fusion, crucially depends on the symmetry and stability of the plasma implosion. Therefore, one of the objectives of the present research was to study the effects of the preembedded axial magnetic field on the plasma instabilities during implosion.

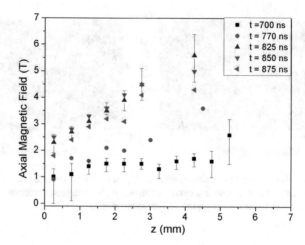

**Fig. 3.12** $B_z$ as a function of $z$ measured on the cylinder symmetry axis at different times of the plasma evolution for $B_{z0} = 0.4\,\mathrm{T}$

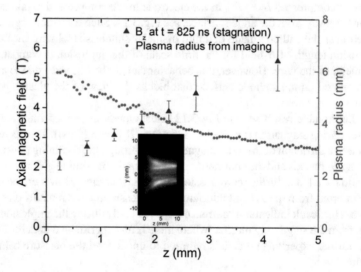

**Fig. 3.13** $B_z$ measured on the cylinder symmetry axis (blue triangles) as a function of $z$ and plasma radius (red dots) as a function of $z$ at $t = 825$ ns for $B_{z0} = 0.4\,\mathrm{T}$

The effects of the axial magnetic field on the plasma instabilities were studied using temporally and spatially resolved 2D visible (4000–6000 Å) imaging, interferometry, and spectroscopy for four different cases of initial axial magnetic-field magnitudes $B_{z0} = 0, 0.1, 0.2,$ and 0.4 T.

Figure 3.15 shows the 2D images of plasma self emission obtained for the four different initial $B_{z0}$ in three different stages of implosion: initial (left column), intermediate (center column), and stagnation (right column). Atomic physics calculations

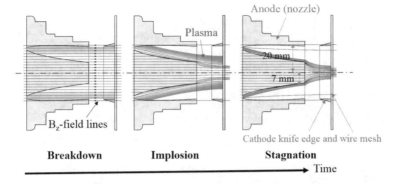

**Fig. 3.14**  Schematic explanation of the electrode effect on the plasma dynamics and $B_z$ distribution

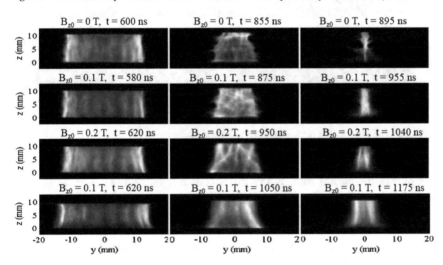

**Fig. 3.15**  2D images of the plasma self emission for different times and initial axial $B$-fields. The cathode mesh and the gas valve anode are, respectively, at the top and the bottom of each image. Left and middle columns show data recorded during inward acceleration

show that the light intensity in the images is roughly proportional to $n_e^2$ integrated along the line of sight and has a week dependence on $T_e$. Figure 3.16 is similar to Fig. 3.15, but for the interferometric images. The deviation of the interference fringes from a straight line is proportional to the electron density integrated along the laser path.

Both, 2D images and interferograms clearly show the development of non-uniformities in the imploding plasma.

For low initial axial magnetic fields ($B_{z0} \leq 0.1$ T) and times t > 700 ns, the plasma column develops distinct cusps at its outer radius. This phenomenon is caused by a magneto-Rayleigh-Taylor instability (MRTI), where the azimuthal magnetic field

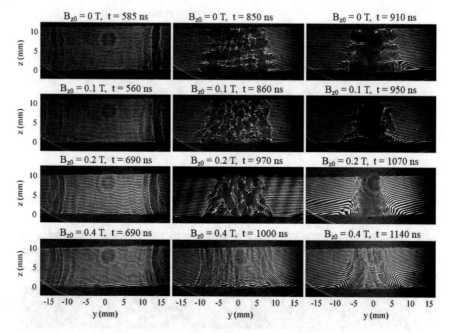

**Fig. 3.16** The same as Fig. 3.15, with interferometric images

that pushes the plasma radially inward, acts as a light fluid [3]. A more rigorous analysis of the MRTI will be given in the following section (Sect. 3.3.1).

For $B_{z0} = 0.2$ T, another phenomenon of distinct filament-like structures is observed, as can be seen in the 2D images in the intermediate stage of compression. These filaments are inclined to the pinch axis, probably following the combined axial and azimuthal magnetic field direction. Several theoretical and experimental works (see Ref. [3] and references therein) also describe similar filamentation phenomena in Z-pinch implosions but without preembedded $B_z$. All of these studies agree on explaining the filament structures as a thermal type instability, but differ in describing the plasma conditions in which they develop. Therefore, in Sect. 3.3.2 we determine $n_e$ and $T_e$ of the filaments and of the surrounding plasma.

All the measurements for studying the instabilities are performed for a relatively high argon gas load of 30 $\mu$g/cm that allowed for maximizing the energy coupling from the current generator to the plasma.

### 3.3.1   Investigation of Magneto-Rayleigh-Taylor Instabilities

The most effective type of instability which can develop during the plasma shell inward acceleration is the magneto-Rayleigh-Taylor instability (MRTI) [3]. There-fore, mitigation of this type of instability is essential for Z-pinch performance

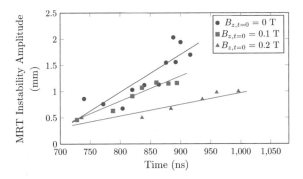

**Fig. 3.17** MRTI amplitudes as a function of time for $B_{z0} = 0$ (blue circles), $B_{z0} = 0.1$ T (green triangles), and $B_{z0} = 0.2$ T (red squares). The corresponding lines are the least squares linear fits that show the amplitude tendency for each case, highlighting the mitigation of MRTI for larger $B_{z0}$

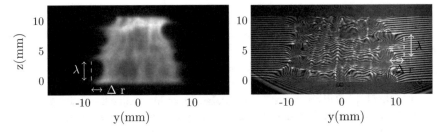

**Fig. 3.18** Illustration of the MRTI amplitude ($\Delta r$) and wavelength ($\lambda$) determination for a 2D image recorded at $t = 855$ ns and for an interferogram recorded at $t = 850$ ns with $B_{z0} = 0$

optimization. There are several theoretical works (e.g. [5]) and experimental observations (e.g. [4]) that, investigated the mitigation of the MRTI by introducing an axial magnetic field. However, most of the measurements in these experiments were performed in the stagnation stage. Experimental data on the perturbation development during the implosion of a Z-pinch in the presence of axial magnetic field are still missing. Therefore, here we investigate the axial magnetic field effects on MRTI during the entire implosion.

Here, the conditions in which the MRTI are most clearly observed are in the plasma column edges for $B_{z0} = 0$ at times >700 ns (Figs. 3.15 and 3.16, upper panels of the middle and right columns). Comparing the images obtained for $B_{z0} = 0$ with those for $B_{z0} > 0$ provides strong evidence for the stabilizing effect of the preembedded axial magnetic field on the implosion. This is demonstrated in Fig. 3.17 that shows the average amplitudes of the MRTI as a function of time for three different $B_{z0}$. The amplitudes are determined from the 2D images recorded at different times of the implosion and are defined as the difference between the radii of adjacent cusp and well, as illustrated in Fig. 3.18. It is seen that the larger $B_{z0}$ is, the smaller are the amplitudes of the MRTI.

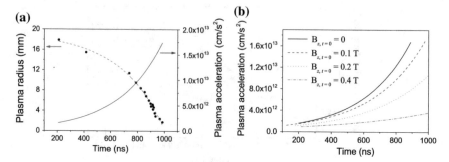

**Fig. 3.19** **a** Time evolution of the plasma radius $R(t)$ for $B_{z0} = 0.1$ T (black dots) and the corresponding fit (red dashed curve) together with the inferred plasma acceleration $\ddot{R}(t)$ (blue curve). **b** Plasma radial acceleration for different $B_{z0}$

An estimate of the MRTI growth rate $\gamma$ for different $B_{z0}$ is obtained using the plasma radial acceleration $g$ and the characteristic wave number of the observed perturbations $k$ [3, 6]:

$$\gamma(t) \approx \sqrt{g(t)\,k} \qquad (3.2)$$

We note that Eq. 3.2 diverges for $k \longrightarrow \infty$. However, in practice the observed fastest growing wavenumber is finite, because some physical processes, which become important for large $k$, are not considered in the derivation of Eq. 3.2 [6].

The plasma acceleration is obtained from the 2D-images by fitting the time evolution of the plasma column radii with an exponentially decaying function: $R(t) = A \exp(-t/\tau) + R_0$ (where $A$, $\tau$ and $R_0$ are fitting parameters), as demonstrated in Fig. 3.19a for $B_{z0} = 0.1$ T. Figure 3.19b presents a comparison of the plasma accelerations as a function of time for different $B_{z0}$. During intermediate implosion times the plasma acceleration is smaller for larger $B_{z0}$, which according to Eq. 3.2 should result in a lower growth rate of the MRTI. This result is consistent with the observation summarized in Fig. 3.17.

The characteristic wavelength $\lambda$ of the MRTI is defined as the average distance between successive cusps seen in the 2D and interferometric images, as shown in Fig. 3.18. It is interesting to note that while the instability amplitudes increase with time the wavelengths don't change significantly. In Table 3.1 the observed characteristic wavelengths and corresponding wavenumbers of the perturbations for different $B_{z0}$ are presented in columns 2 and 3, respectively (For $B_{z0} = 0.4$ T, no clear MRTI are observed). It is clearly seen that the observed perturbation wavenumber decreases for increasing $B_{z0}$. This suggests that the restoring force of the bending axial magnetic field-lines significantly affects the wavenumber of the fastest growing MRTI mode, and therefore impedes the growth of MRTI.

The knowledge of the perturbation wavenumber and the plasma acceleration allows for estimating the total perturbation amplitude growth due to MRTI [3]. The calculation is based on the instantaneous growth rate ($\sqrt{g(t)k}$) integrated over the implosion time (using the WKB approximation):

**Table 3.1** Parameters for the MRTI study for different $B_{z0}$ at intermediate times of implosion: (1) $B_{z0}$ magnitudes, (2) wavelengths of the observed MRT perturbations and (3) corresponding wavenumbers of the MRT perturbations, (4) stagnation time, and (5) number of e-foldings (Eq. 3.3)

| $B_{z0}$ T | $\lambda$ mm | $k = 2\pi/\lambda$ mm$^{-1}$ | $t_{stag}$ ns | Number of e-foldings $\int_{t_0}^{t_{stag}} \gamma(t')\, dt'$ |
|---|---|---|---|---|
| 0 | 2 | 3.1 | 890 | 7 |
| 0.1 | 5 | 1.3 | 940 | 5.5 |
| 0.2 | 5 | 1.3 | 1020 | 4.5 |
| 0.4 | >10 | <0.6 | 1160 | 2 |

$$A(t) \propto \exp\left(\int_{t_0}^{t_{stag}} \gamma(t')\, dt'\right) = \exp\left(\int_{t_0}^{t_{stag}} \sqrt{g(t')\, k}\, dt'\right) \quad . \tag{3.3}$$

Here, $t_0$ denotes the time when the instability starts to grow, determined by the first appearance in the 2D images and interferograms, and $t_{stag}$ is the stagnation time. The plasma stagnation typically coincides with a sharp maximum in UV radiation and was determined by a UV (1200–2500 Å) photo diode. The results of the integral in Eq. 3.3 (number of e-foldings) for the different $B_{z0}$ are given in the last column of Table 3.1.

It is seen from Table 3.1 that larger $B_{z0}$ lead to a lower number of e-foldings, and thus to reduced MRTI growth. In principle, smaller perturbation wavenumbers do not necessarily reduce the MRTI growth since they are accompanied by longer implosion times until stagnation (see Table 1, column 6). However, the increase in the implosion time with $B_{z0}$ is less significant than the decrease of the other parameters in Eq. 3.3, thus the MRTI growth is reduced.

Quantitatively, the number of e-foldings is decreased by 20% for $B_{z0} = 0.1$ T and by 35% for $B_{z0} = 0.2$ T (in comparison to $B_{z0} = 0$). This result can be related directly to the measured MRTI amplitudes (see Fig. 3.17). For $B_{z0} = 0$ the observed amplitude at the stagnation time is $\sim$2 mm and decreases by 20% to $\sim$1.6 mm for $B_{z0} = 0.1$ T. For $B_{z0} = 0.2$ T the MRTI amplitude is $\sim$1 mm representing a decrease by 50%. It should be noted that the calculated number of e-foldings for $B_{z0} > 0$ constitutes an upper limit since the calculation is based on a theory that does not specifically consider the effect of the $B_z$-field lines bending. The effect of $B_z$ is taken into account only indirectly through the observed wavenumber and acceleration.

## 3.3.2 Investigation of Filament-Like Structures in Implosions with Axial Magnetic Field

Filtered 2D images (bandpass filter 4000–6000 Å) of the plasma recorded at intermediate times of implosion reveal the existence of regions in the plasma, in a form of

68

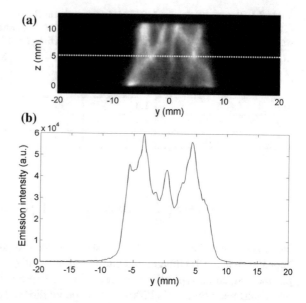

**Fig. 3.20** **a** 2D image of the visible plasma self emission recorded at $t = 960$ ns for $B_{z0} = 0.2$ T, and **b** lineout of the 2D image taken at $z = 5$ mm

filaments, with significantly different emission intensity. The differences are especially pronounced in implosions with $B_{z0} = 0.2$ T, as can be seen in Fig. 3.20.

Figure 3.20a presents the 2D image of the plasma column recorded during the implosion at $t = 960$ ns which demonstrates the existence of filament-like structures (regions of higher emission) that are inclined to the $z$-axis. As seen in the lineout (Fig. 3.20b), taken at $z = 5$ mm (marked by the dotted line in Fig. 3.20a), the emission intensity of the filaments in the visible spectral region is up to 50% higher than that of the surrounding plasma. Our conjecture is that the filaments lye along the direction of the combined $B_z$ and $B_\theta$ fields. Assuming this is true, this indicates that $B_z$ and $B_\theta$ are comparable also in implosion with $B_{z0} = 0.2$ T and mass per length of 30 μg/cm (the measurements presented in Sect. 3.1 are for $B_{z0} = 0.4$ T and mass per length of 10 μg/cm). This also means that a large part of the current doesn't flow through the imploding plasma. Rather the current flows outside, since from the measured total current and observed plasma radius, the expected $B_\theta$ is ∼9 T and it is significantly larger than the calculated maximum $B_z$ (∼3 T) obtained by assuming flux conservation.

Time-resolved imaging spectroscopy (see Fig. 2.6) was used to study the plasma parameters of both entities, the filaments and the surrounding plasma. The measurements were performed for initial conditions of $B_{z0} = 0.2$ T, at times of 900–1000 ns from the beginning of the implosion, when the filaments phenomenon is found to be most pronounced. Figure 3.21 presents two examples of argon plasma spectral images recorded at $t = 960$ ns and $t = 1000$ ns, at $z = 5$ mm. These measurements provide the emission intensity integrated along the line of the radial observation. Generally, for cylindrically symmetric z-pinch sources, an inverse Abel transform can be performed to obtain the radial distribution of the emission. However, in the

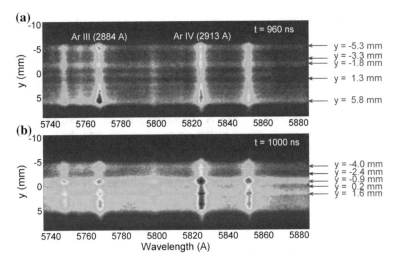

**Fig. 3.21** **a** Spectroscopic image of argon plasma emission at $z = 5$ mm for $B_{z0} = 0.2$ T, recorded at $t = 960$ ns. **b** The same as **a**, but for $t = 1000$ ns. The second order of the Ar III and Ar IV UV-lines is recorded. The arrows along with the y coordinates show the y-positions where lineouts for detailed analysis are generated. The blue and red arrows represent low and high intensity regions, respectively

present case, the filaments disturb the cylindrical symmetry making the inverse Abel transform unreliable.

The spectral region of the measurements is chosen to consist emission lines of two argon ion charge states, Ar III and Ar IV. This allows for electron temperature and density determination from a single measurement. It is noted that the second order of the Ar III and Ar IV UV-lines is recorded, since the spectrometer grating has a higher efficiency and provides better spectral resolution in the second order. It was verified that no first order emission lines are present in the measured spectral window. The only contribution of the first order emission comes from continuum radiation.

To determine the plasma parameters in the different regions, we analyze lineouts obtained from plasma regions with significantly different emission intensities (see Fig. 3.21). At electron densities relevant to this analysis ($>10^{18}$ cm$^{-3}$), the emission line shapes are dominated by Stark broadening ($1.7 \leq FWHM_{Stark} \leq 2.6$ Å), whereas the contributions of the instrumental ($\sim0.5$ Å) and Doppler (due to thermal and hydrodynamic motion) ($<0.3$ Å) broadenings are relatively small. Therefore, the emission lines in the measured spectra are fitted using a Lorentz distribution.

Figure 3.22 presents the lineout at $y = 0.2$ mm for the spectral image recorded at $t = 1000$ ns along with the best fit, obtained by varying the parameters of five Lorentz distributions, corresponding to the transitions seen at $\lambda = 2874$, 2884, 2899, 2913, and 2926 Å. The feature seen at $\lambda = 2899$ Å probably arises from the blending of several unidentified weak transitions. It is approximated by a single, wide Lorentzian, whose wings contribute very little to the other transitions and therefore do not disturb their diagnostics.

The full width at half maximum (FWHM) of each fitted Lorentzian defines the Stark broadening of the line that is used for the $n_e$ determination. The area of each fitted Lorentzian gives the line intensity that is used for the $T_e$ determination. The Ar III $(2D^o)4s\,^3D^o_3-(2D^o)4p\,^3P_2$ transition at $\lambda = 2884$ Å and Ar IV $(3P)4s\,^2P_{3/2}-(3P)4p\,^2D^o_{5/2}$ transition at $\lambda = 2913$ Å are selected for the detailed analysis due to their highest emission intensity in the observed region, and the availability of theoretical and experimental data on their Stark broadening. These transitions are isolated and belong to non-hydrogenlike ions. Therefore, the relation between the FWHM and $n_e$ is linear, with proportionality constant found in [7–9] and is given by (Eqs. 1.50 and 1.51 in Sect. 1.3.1):

$$\Delta\lambda^{2884}_{Stark}(\text{Å}) = 1.1 \times 10^{-18} \cdot n_e(\text{cm}^{-3})$$
$$\Delta\lambda^{2913}_{Stark}(\text{Å}) = 0.7 \times 10^{-18} \cdot n_e(\text{cm}^{-3})$$

$\Delta\lambda_{Stark}$ is the full width at half maximum of the Lorentzian distribution. The theoretical transition intensity ratios as a function of $T_e$ for constant densities are calculated using a collisional-radiative (CR) code [10] and presented in Fig. 1.10.

In Fig. 3.23, the normalized spectral lineouts, obtained from Fig. 3.21b at $y = 0.2$ mm (corresponding to the low intensity, surrounding plasma region) and at $y = -0.9$ mm (corresponding to a filament) are shown together in order to demonstrate the difference in the lineshapes. The spectrum obtained from the filament have larger width than the lines from the ambient plasma, indicating a higher average $n_e$ along the chord at $y = -0.9$ mm. Another feature observed in Fig. 3.23 is that the line intensity ratio Ar IV/Ar III is larger for the spectrum obtained at $y = 0.2$ mm indicating a higher average $T_e$ in the surrounding plasma. The electron densities and temperatures obtained from the lineouts taken at the different y-positions (see Fig. 3.21) are presented in Table 3.2. The other observed two transitions at $\lambda = 2874$ and $\lambda = 2926$ Å both belong to Ar IV and their analysis gives results that are consistent with the Ar IV transition at $\lambda = 2913$ Å. The $n_e$ inferred from these lines are within 15% of the $n_e$ obtained from the $\lambda = 2913$ Å transition, and their relative intensity is constant, therefore they don't provide additional information on the temperature.

The error estimate for $n_e$ and $T_e$ consists of two contributions: the first, accounts for the uncertainty in the measured parameters of the line intensities and FWHM's, and the second accounts for the uncertainty in the theoretical calculations and published data. For all the measured parameters the error is $\sim$15%. The uncertainty in the data of the Stark broadening for both lines is $\sim$40%. Therefore, the total estimated error for the $n_e$ determination from both lines is $\sim$55%. The uncertainty in the CR model results used for the $T_e$ determination is commonly estimated to be $\sim$20%. Another contribution to the error in the $T_e$ determination results from the uncertainty in the $n_e$ used as an input for the CR calculations. These contributions result in a total uncertainty of $\sim$30% in $T_e$.

**Table 3.2** Summary of plasma parameters obtained from regions with low emission intensity and high emission intensity

| y-Position mm | $FWHM_{ArIII}$ Å | $n_{e,ArIII}$ (cm$^{-3}$) | $FWHM_{ArIV}$ Å | $n_{e,ArIV}$ cm$^{-3}$ | Intensity Ratio Ar III/Ar IV | $T_e$ eV | Comments |
|---|---|---|---|---|---|---|---|
| $t = 960$ ns | | | | | | | |
| **−5.3** | 2.1 | $1.9 \times 10^{18}$ | 2.3 | $3.3 \times 10^{18}$ | 1.1 | 4.7 | Filament |
| **−3.3** | 1.8 | $1.6 \times 10^{18}$ | 1.8 | $2.5 \times 10^{18}$ | 0.8 | 4.75 | Ambient |
| **−1.8** | 2.1 | $1.9 \times 10^{18}$ | 1.9 | $2.7 \times 10^{18}$ | 1.1 | 4.65 | Filament |
| **1.3** | 1.9 | $1.7 \times 10^{18}$ | 1.7 | $2.4 \times 10^{18}$ | 0.8 | 4.75 | Ambient |
| **5.8** | 2.1 | $1.9 \times 10^{18}$ | 2.3 | $3.3 \times 10^{18}$ | 1.3 | 4.6 | Filament |
| $t = 1000$ ns | | | | | | | |
| **−4.0** | 2.4 | $2.2 \times 10^{18}$ | 2.6 | $3.7 \times 10^{18}$ | 0.7 | 5 | Filament |
| **−2.4** | 2.3 | $2.1 \times 10^{18}$ | 2 | $2.9 \times 10^{18}$ | 0.4 | 5.2 | Ambient |
| **−0.9** | 2.6 | $2.4 \times 10^{18}$ | 2.6 | $3.7 \times 10^{18}$ | 0.5 | 5.2 | Filament |
| **0.2** | 2.2 | $2 \times 10^{18}$ | 2 | $2.9 \times 10^{18}$ | 0.4 | 5.2 | Ambient |
| **1.6** | 2.4 | $2.2 \times 10^{18}$ | 2.3 | $3.3 \times 10^{18}$ | 0.5 | 5.15 | Filament |

**Fig. 3.22** Lineout at $y = 0.2$ mm of the spectroscopic image recorded at $t = 1000$ ns, together with the best fit obtained by varying the parameters of five independent Lorentzians. We note that the second order of Ar III and Ar IV lines are recorded

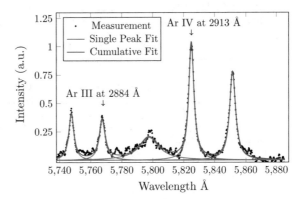

Table 3.2 shows that, despite the rather large uncertainty in the absolute $n_e$ values, the trend of higher densities by ∼10–20% inside the filament region in comparison to the ambient plasma, is clear. The $T_e$ analysis shows that the filaments are regions with slightly lower temperature than the ambient plasma. We note that due to the integration along the line of sight for the filament regions (the lineouts of the filaments might include also contributions from the ambient plasma), the differences in $n_e$ and $T_e$ between the two plasma regions are even larger than those in Table 3.2. Therefore, $n_e$ and $T_e$ presented in Table 3.2 for filaments constitute, respectively, lower and upper limits.

**Fig. 3.23** Comparison of the lineouts at $y = 0.2$ mm (blue curve) and at $y = -0.9$ mm (red curve, filament) generated from the spectroscopic image recorded at $t = 1000$ ns. We note that the second order of Ar III and Ar IV lines are recorded

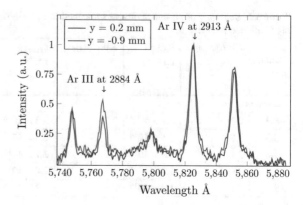

# References

1. Davara G, Gregorian L, Kroupp E, Maron Y (1998) Spectroscopic determination of the magnetic field distribution in an imploding plasma. Phys Plasmas 5(1):1068–1075
2. Mikitchuk D, Cvejic M, Doron R, Kroupp E, Stollberg C, Maron Y, Velikovich AL, Ouart ND, Giuliani JL, Mehlhorn TA, Yu EP, Fruchtman A (2019) Effects of a preembedded axial magnetic field on the current distribution in a Z-pinch implosion. Phys Rev Lett 122:045001
3. Liberman MA, Spielman RB, Toor A, Groot JS (1999) Physics of high-density Z-pinch plasmas. Springer
4. Felber FS, Wessel FJ, Wild NC, Rahman HU, Fisher A, Fowler CM, Liberman MA, Velikovich AL (1988) Ultrahigh magnetic fields produced in a gas-puff z pinch. J Appl Phys 64(8):3831–3844
5. Budko AB, Felber FS, Kleev AI, Liberman MA, Velikovich AL (1989) Stability analysis of dynamic z pinches and theta pinches. Phys Fluids B Plasma Phys 1(3):598–607
6. Ryutov DD et al (2000) The physics of fast Z pinches. Rev Mod Phys 72:167
7. Konjević N, Lesage A, Fuhr JR, Wiese WL (2002) Experimental stark widths and shifts for spectral lines of neutral and ionized atoms (a critical review of selected data for the period 1989 through 2000). J Phys Chem Ref Data 31(3):819–927
8. Sahal-Brehot S, Dimitrijevic MS, Moreau N (August 2015) Stark-b database
9. Hamdi R, Nessib NB, Sahal-Brechot S, Dimitrijevic MS (2014) Stark widths of Ar III spectral lines in the atmospheres of subdwarf B stars. Adv Space Res 54(7):1223–1230. Spectral line shapes in astrophysics and related phenomena
10. Ralchenko YuV, Maron Y (2001) Accelerated recombination due to resonant deexcitation of metastable states. J Quant Spectr Rad Transfer 71:609

# Chapter 4
# Discussion

The most significant result of this study is the first direct measurement of the evolution of the current distribution in Z-pinch with preembedded axial magnetic field ($B_{z0}$). This result was achieved by measuring the azimuthal magnetic field ($B_\theta$) in the imploding plasma. In addition, the evolution and distribution of the axial magnetic field that is compressed by an imploding plasma is measured simultaneously with the compressing $B_\theta$-field. These measurements revealed unexpected plasma dynamics in Z-pinch with preembedded $B_{z0}$. While without the application of an initial $B_z$ the measurements show, as expected, that nearly the whole current flows through the imploding plasma (see Fig. 3.7), when an initial $B_{z0} = 0.4$ T is applied the measured $B_\theta$ in the imploding plasma is much smaller than the expected value (calculated from the total current and plasma radius, see Fig. 3.8), showing that only a small part of the current flows in the imploding plasma. Measurements of $B_\theta$ at larger radii show that the missing current flows at radii much larger than the outer radius of the imploding plasma. Moreover, for the $B_{z0} = 0.4$ T case, the value of $B_\theta$ at the outer radius of the imploding plasma remains nearly constant (between 1.5 and 2 T) during the implosion, indicating a decrease of current within the imploding plasma as the implosion progresses. This loss of current explains two observations: (i) the significantly longer implosion time and (ii) much lower $B_{z,stagnation}$ than the values predicted theoretically for $B_{z0} = 0.4$ T, see Sect. 1.2.1 and Table 1.4. This findings sheds light on several unexplained phenomena observed in other Z-pinch experiments with preembedded axial magnetic field: (i) the formation of helical structures [1, 2], (ii) larger than predicted implosion time and plasma radius at stagnation [3, 4], (iii) stronger than predicted mitigation of instabilities [1, 2, 5–7], and (iii) unexpectedly strong reduction of the K-shell emission at relatively small $B_{z0}$ [5].

Figures 4.1 and 4.2 present the measured plasma radius and $B_z$ as a function of time for an implosion with $B_{z0} = 0.4$ T. These measurements are compared to different calculations based on the theory described in Sect. 1.2.1. The red curves represent calculations of the plasma radius (Fig. 4.1) and the $B_z$ (Fig. 4.2) evolution

© Springer Nature Switzerland AG 2019
D. Mikitchuk, *Investigation of the Compression of Magnetized Plasma and Magnetic Flux*, Springer Theses,
https://doi.org/10.1007/978-3-030-20855-4_4

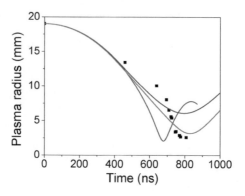

**Fig. 4.1** Plasma radius as a function of time for $B_{z0} = 0.4$ T. Black squares are the experimental results (the plasma radius is defined as the radius of maximum emission intensity after inverse Abel transform); red solid line—theoretical calculation assuming the whole current flows through the imploding plasma and conservation of $B_z$-flux; blue solid line—theoretical calculation assuming constant $B_\theta = 1.6$ T and conservation of $B_z$-flux; green solid line—theoretical calculation assuming constant $B_\theta = 1.6$ T and $B_{z0} = 0.2$ T (to consider the effect of diffusion of $B_{z0} = 0.4$ T.)

for $B_{z0} = 0.4$ T, assuming the whole current flows through the imploding plasma and conservation of $B_z$-flux. The blue curves represent similar calculations, but assuming that the time evolution of the current flowing through the imploding plasma changes such that the compressing $B_\theta$ remains constant and equals to 1.6 T that is the average of the measured $B_\theta$ values. The green curves represent similar calculations with a current evolution that maintains $B_\theta = 1.6$ T, but with an additional effect of $B_z$ diffusion. The effect of the diffusion of $B_z$ out of the plasma is included by artificially reducing the magnitude of the initial $B_z$ by 50% ($B_{z0} = 0.2$ T), according to the measured confinement efficiency of the plasma. Comparing the theoretical calculations with the experimental results we see that the red curves differ significantly from the measurements: the calculated implosion time is $\sim 14\%$ shorter and $B_{z,stagnation}$ is more than $5\times$ larger then observed. When a constant $B_\theta$ of 1.6 T is assumed and $B_{z0} = 0.4$ T, represented by the blue curves, the calculated implosion time is improved, but the predicted plasma radius at stagnation is $2\times$ larger than the measured value, and the predicted $B_{z,stagnation}$ is $\sim 2$ times smaller than the measured value. In the third model, where the effect of $B_z$ diffusion is considered, represented by the green curves, both, the predicted implosion time and radius at stagnation are closer to the measured values, as well as the evolution of $B_z$ near stagnation, supporting the choice of the model. However, also this latter model fails to accurately describe the measured plasma radius evolution during the implosion stage. This difference suggests that the simplified theory used here to describe the plasma dynamics and the evolution of $B_z$ is not sufficiently accurate. A better description would requires a much more complicated model that will take into account the $B_z$ diffusion and modifications to the current flowing through the imploding plasma.

The measurements of $B_\theta$, $B_z$, and $t_{stagnation}$ show that the application of initial $B_z$ significantly affects the current distribution in the plasma, such that most of the

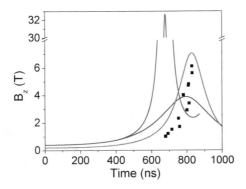

**Fig. 4.2** $B_z$ as a function of time for $B_{z0} = 0.4$ T. Black squares are the experimental results ($B_z$ is measured at $z = 3$ mm on the symmetry axis); red solid line—theoretical calculation assuming whole current flows through the imploding plasma and conservation of $B_z$-flux; blue solid line—theoretical calculations assuming constant $B_\theta = 1.6$ T and conservation of $B_z$-flux; green solid line—theoretical calculations assuming constant $B_\theta = 1.6$ T and $B_{z0} = 0.2$ T (to consider the effect of diffusion of $B_{z0} = 0.4$ T

current flows at radii larger than the radius of the imploding main plasma. Recent measurements [3] (not presented here) show the existence of a low-density peripheral plasma ($10^{16} < n_e < 10^{17}$ cm$^{-3}$ compared to $\sim 10^{18}$ cm$^{-3}$ of the dense imploding plasma) residing at $20 < r < 28$ mm that carries a significant part of the current. We found a two sources for this low-density peripheral plasma. The first source is the gas-load atoms that expanded to radii larger than $R_0 = 19$ mm either due to the backscattering from the cathode or due to the expansion of the boundary layers, and are not swept by the main plasma implosion. The second source is atoms (mainly hydrogen, carbon and oxygen atoms) that were absorbed on the electrodes prior to the discharge or emitted from the plastic housing of the Helmholtz coils during the discharge. This low-density plasma at large radii exists also when the initial $B_z$ is not applied. However, when $B_{z0} = 0$ it has lower electron density and temperature ($n_e < 10^{16}$ cm$^{-3}$, $T_e < 1.5$ eV), consistent with the absence of the current there.

To explain this phenomenon of the current re-distribution from the dense imploding plasma to the low-density peripheral plasma when external $B_z$ is applied, we propose a simplified model based on the development of a force-free current in the low-density plasma. In the model we employ planar geometry and consider plasma in constant and uniform external electric $\vec{E} = E_z \hat{z}$, and magnetic $\vec{B} = B_\theta \hat{\theta} + B_z \hat{z}$ fields. Note, that although the considered geometry is planar, for convenience we use cylindrical coordinates notation, whereas the equivalence between the coordinate systems is the following: $\hat{r} = \hat{x}$, $\hat{\theta} = \hat{y}$, and $\hat{z} = \hat{z}$. To find the evolution of the plasma velocity and the current density we solve a simplified version of the equation of motion Eq. 1.10, and the generalized Ohm law Eq. 1.4:

$$\frac{d\vec{v}}{dt} = \frac{\vec{j} \times \vec{B}}{\rho} \tag{4.1}$$

$$\frac{d\vec{j}}{dt} = \frac{\nu_{ei}}{\eta}\left(\vec{E} + \vec{v} \times \vec{B}\right) - \nu_{ei}\vec{j} - \omega_{ce}\frac{\vec{j} \times \vec{B}}{|\vec{B}|} \tag{4.2}$$

with initial conditions: $\vec{v}(t = 0) = 0$ and $\vec{j}(t = 0) = 0$, $\vec{v}$ is the plasma velocity, $\vec{j}$ is the current density, $\rho$ is the plasma mass density, $\nu_{ei}$ is electron-ion collision frequency, $\eta$ is the plasma resistivity, and $\omega_{ce}$ is the electron cyclotron frequency. In Eqs. 4.1 and 4.2 the pressure terms are omitted since in the low-density plasma residing at large radii the magnetic pressure is much larger than the thermal pressure ($\beta \sim 0.1$) and the spatial gradients of the density and temperature are small (see [3]).

Equations 4.1 and 4.2 can be solved analytically as shown in Appendix A.1. This solution shows that the current density and plasma velocity reach steady state values given by:

$$j_z = \frac{E_z}{\eta}\frac{B_z^2}{B_z^2 + B_\theta^2} \qquad j_\theta = \frac{E_z}{\eta}\frac{B_z B_\theta}{B_z^2 + B_\theta^2} \qquad \vec{v} = \frac{\vec{E} \times \vec{B}}{|\vec{B}|^2} = \frac{E_z B_\theta}{B_z^2 + B_\theta^2}\hat{r} \tag{4.3}$$

on a time-scale:

$$\tau_{steady} = \frac{\nu_{ei}}{\omega_{ce}\omega_{ci}} \tag{4.4}$$

Using the plasma parameters $10^{16} \leq n_e \leq 10^{17}$ cm$^{-3}$ and $T_e \sim 5$ eV, and the magnetic field $B \approx 2$ T, typical for the low-density plasma [3], $\tau_{steady}$ is $\sim 15$ ns. This time is much shorter than the implosion time (few hundred ns). Therefore, the steady state solution given by Eq. 4.3 might be valid in the low-density plasma. An interesting prediction of this solution is a development of an azimuthal current density $j_\theta$ in the low-density plasma. This current might generate an additional $B_z$-flux in the region of the dense imploding plasma, thereby, affecting the imploding plasma dynamics and the development of instabilities. Further measurements are required to check this prediction of $B_z$ enhancement.

An insightful approach to the steady state solution of the model is to calculate the electric field (using Galileo transformation) in the rest frame of the plasma moving with the drift velocity given in Eq. 4.3:

$$\vec{E}_{pl} = \vec{E}_{lab} + \vec{v} \times \vec{B} = \vec{E}_{lab} + \frac{\vec{E}_{lab} \times \vec{B}}{|\vec{B}|^2} \times \vec{B} = \vec{E}_{lab} + \frac{1}{|\vec{B}|^2}((\vec{E}_{lab} \cdot \vec{B})\vec{B} - |\vec{B}|^2\vec{E}_{lab}) =$$

$$= \frac{(\vec{E}_{lab} \cdot \vec{B})\vec{B}}{|\vec{B}|^2} \tag{4.5}$$

where $\vec{E}_{pl}$ and $\vec{E}_{lab}$ are the electric fields in the rest frame of the plasma and laboratory, respectively. From Eq. 4.5 we see that the electric field in the rest frame of plasma is zero when $\vec{E}_{lab}$ is perpendicular to $\vec{B}$ as in the classical Z-pinch configuration. However, when external axial magnetic field is applied, $\vec{E}_{pl} = (E_{z,lab} \cdot B_z)\vec{B}/|\vec{B}|^2 \neq 0$.

Using generalized Ohm's law (Eq. 1.4) we see that for the case $\vec{E}_{lab} \perp \vec{B}$, the current in a plasma moving at the drift velocity can be driven only by spatial gradients in the plasma parameters. An estimate of the current in the low-density plasma due to the spatial gradients in $n_e$ and $T_e$ shows that it is very small compared to the total current, therefore, when $B_{z0} = 0$ the entire current can flow in the imploding dense plasma. On the other hand, if $B_{z0} > 0$, then $\vec{E}_{pl} \neq 0$, allowing current to flow in the low-density plasma with current densities given by Eq. 4.3.

Another observation we discuss here in the context of this model is the slow or non implosion of the low-density plasma, although it carries up to $\sim$50–80% of the total current. One would expect that such current should accelerate the low-density plasma to relatively large implosion velocity due to the Lorentz force ($\vec{j} \times \vec{B}$) directed inward. The question is, whether the implosion time of the low-density plasma is large or small compared to the implosion time of the dense plasma (few hundreds ns). If it is small, then the low-density plasma that carries part of the current would reach the dense plasma in contradiction to our observation. To estimate the inward motion of the low-density plasma we use radial velocity equation 4.3 that gives $v_r \sim 7.5 \times 10^5$ cm/s. For calculating $v_r$, we estimate $E_z$ using the expression for $j_z$ of Eq. 4.3 and assuming a uniform current distribution in $20 < r < 28$ mm, and Spitzer's resistivity (Eq. 1.5). We also used $B_z \sim B_\theta \sim 1$, here $B_z > B_{z0}$ due to possible enhancement of $B_z$ by $j_\theta$ flow in the low-density plasma. The estimated $v_r \sim 7.5 \times 10^5$ cm/s of the low-density plasma is much smaller than the average implosion velocity $\sim 3 \times 10^6$ cm/s of the dense plasma therefore, consistent with the observation. It is important to point out that even if the presented model is not valid, the estimate of the Lorentz force and the the the resulting implosion of the low-density plasma predicts that on the time scale of the implosion the radial displacement of the low-density plasma is small relatively to its initial radius. For the estimate of the low-density velocity we use the following plasma parameters: $I_{ldp} = 100$ kA, $20 \leq r_{ldp} \leq 28$ mm, and $n_{i,carbon} \approx 3 \times 10^{16}$ cm$^{-3}$. Assuming a uniform current distribution, the current density in the low-density plasma is $j_{ldp} \approx 8 \times 10^7$ A/m$^2$. The average $B_\theta$ in the low-density plasma for the total current $\sim$200 kA is $\sim$1.5 T. Therefore, the inward acceleration is:

$$a_{ldp} \sim \frac{j_{ldp} \cdot B_\theta}{n_{i,carbon} \cdot m_C} = 2 \times 10^{11} \ (\text{m/s}^2) \tag{4.6}$$

where $m_C$ is the mass of a carbon atom. The radial displacement of the low-density plasma in 100 ns is $\Delta r \sim at^2/2 = 1$ mm. Therefore, the low-density plasma will implode by $\sim$4 mm during the time of the current re-distribution ($\sim$400 ns before the stagnation). This radial displacement is small compared to the initial radius of the low-density plasma ($\sim$25 mm), consistent with the observation. In future experiments we plan to measure the current distribution in the low-density plasma in order to verify our estimate.

It is important to note that this simplified model is not applicable to the dense imploding plasma due to following three reasons: (i) the magnetic field is mostly self generated and not external; (ii) assumption of constant $\vec{E}$ and $\vec{B}$ is not valid since, for the plasma parameters and magnetic field of the imploding plasma, $\tau_{steady} \sim t_{implosion}$; and (iii) the pressure terms in the Eqs. 1.10 and 1.4 can not be neglected.

We point out that the model discussed here does not include Maxwell's equations, i.e. the $B$-field diffusion processes (through Faraday's law, Eq. 1.2) and the self fields (through Ampere's law, Eq. 1.3). Although, the neglect of these processes is not justified for the parameters of the low-density plasma, the tendency of the plasma to establish force-free configurations will not change with the inclusion of the Maxwell's equations as shown in [8, 9]. However, the time to reach steady state might be affected. To test the validity of this simplified model for describing the governing physics, three-dimensional modeling involving processes beyond resistive MHD (for example, the Hall term in generalized Ohm's law) is required. Such simulations of the Z-pinch implosion with preembedded axial magnetic field have been recently published [10, 11] that demonstrate the importance of the low-density plasma for the current distribution and the development of instabilities in the imploding plasma.

Besides the model presented above, it is also useful to consider the dynamics of the electrons in constant and uniform electric and magnetic fields without the ion dynamics. This treatment might be relevant for times prior to the force-free configuration establishment (see [12], Chap. 3.1 ).

The planar solution for a collisionless electron in $\vec{E} = (0, 0, E_z)$ and $\vec{B} = (0, B_y, B_z)$ is derived in Appendix A.2 giving the guiding center velocity:

$$\vec{v}_{gc} = -\left( \frac{E_z B_y}{B_y^2 + B_z^2}, \frac{E_z B_z}{B_y^2 + B_z^2} \omega_{ce}^{(y)} \cdot t, \frac{E_z B_z}{B_y^2 + B_z^2} \omega_{ce}^{(z)} \cdot t \right) \tag{4.7}$$

where $\omega_{ce}^{(y)} \equiv e B_y / m_e$, $\omega_{ce}^{(z)} \equiv e B_z / m_e$.

The planar steady state solution for collisional electrons in $\vec{E} = (0, 0, E_z)$ and $\vec{B} = (0, B_y, B_z)$ is derived in Appendix A.2 giving the guiding center velocity:

$$\vec{v}_{gc} = -\left( \frac{\frac{\omega_{ce}^{(y)}}{\nu_{ei}} v_{drift}}{1 + \left( \frac{(\omega_{ce}^{(y)})^2 + (\omega_{ce}^{(z)})^2}{\nu_{ei}^2} \right)}, \frac{\frac{\omega_{ce}^{(y)} \omega_{ce}^{(z)}}{\nu_{ei}^2} v_{drift}}{1 + \left( \frac{(\omega_{ce}^{(y)})^2 + (\omega_{ce}^{(z)})^2}{\nu_{ei}^2} \right)}, \frac{\left( 1 + \left( \frac{\omega_{ce}^{(z)}}{\nu_{ei}} \right)^2 \right) v_{drift}}{1 + \left( \frac{(\omega_{ce}^{(y)})^2 + (\omega_{ce}^{(z)})^2}{\nu_{ei}^2} \right)} \right) \tag{4.8}$$

where $v_{drift} \equiv e E_z / (m_e \nu_{ei})$.

Both solutions show that application of $B_z$ aligned with the external electric field, increases the transport of electrons in the $\hat{z}$-direction. We see from Eq. 4.8 that the parameter defining the electron transport in the problem with collisional electrons is the ratio of the electron gyro-frequency ($\omega_{ce} \approx 1.76 \times 10^{11} \cdot B(T)$) to the electron-ion collision frequency ($\nu_{ei}$, see Eq. 1.6) as explained in Appendix A.2.

To demonstrate the effect of $B_\theta$ on the electrons velocity in $\hat{z}$-direction in the low-density plasma, we estimate this ratio for implosion without axial magnetic field using the following plasma parameters and current [13]: $n_{e,ldp} \sim$ few $\times 10^{15}$ cm$^{-3}$, $T_{e,ldp} \sim 1$ eV, $r_{ldp} \sim 24$ mm, and $I \sim 150$ kA. These parameters give: $B_\theta(r = r_{ldp}) = 1.25$ T, $\omega_{ce} \approx 2.2 \times 10^{11}$ rad/s, and $\nu_{ei} \approx 4 \times 10^{10}$. s$^{-1}$ that result in $ratio \sim$ 5.5. As can be seen from Appendix A.3, for $ratio \gg 1$, $v_{gc,z} \approx ratio^{-2} \cdot v_{drift} \ll v_{drift}$. We see that in implosions without $B_z$, electrons in the low-density plasma are magnetized and according to the considered problem should carry only a small portion of the current perpendicular to $B_\theta$, in comparison to the dense imploding plasma ($ratio \sim 0.1$ leading to $v_{gc,z} \approx v_{drift}$).

# References

1. Mikitchuk D, Stollberg C, Doron R, Kroupp E, Maron Y, Strauss HR, Velikovich AL, Giuliani JL (2014) Mitigation of instabilities in a z-pinch plasma by a preembedded axial magnetic field. Trans Plasma Sci 42(10):2524–2525
2. Awe TJ et al (2013) Observations of modified three-dimensional instability structure for Imploding Z-Pinch liners that are premagnetized with an axial field. Phys Rev Lett 111:235005
3. Mikitchuk D, Cvejic M, Doron R, Kroupp E, Stollberg C, Maron Y, Velikovich AL, Ouart ND, Giuliani JL, Mehlhorn TA, Yu EP, Fruchtman A (2019) Effects of a preembedded axial magnetic field on the current distribution in a Z-pinch implosion. Phys Rev Lett 122:045001
4. Rousskikh AG, Zhigalin AS, Oreshkin VI, Baksht RB (2017) Measuring the compression velocity of a z pinch in an axial magnetic field. Phys Plasmas 24(6):063519
5. Shishlov AV, Baksht RB, Chaikovsky SA, Fedunin AV, Fursov FI, Kokshenev VA, Kurmaev NE, Labetsky AY, Oreshkin VI, Ratakhin NA, Russkikh AG, Shlykhtun SV (2006) Formation of tight plasma pinches and generation of high-power soft x-ray radiation pulses in fast z-pinch implosions. Laser Phys 16(1):183–193
6. Qi N, de Grouchy P, Schrafel PC, Atoyan L, Potter WM, Cahill AD, Gourdain P-A, Greenly JB, Hammer DA, Hoyt CL, Kusse BR, Pikuz SA, Shelkovenko TA (2014) Gas puff z-pinch implosions with external bz field on cobra. AIP Conf Proc 1639(1):51–54
7. Conti F, Valenzuela J, Ross MP, Narkis J, Beg F (2018) Stability measurements of a staged z-pinch with applied axial magnetic field. In: The 45th IEEE International Conference on Plasma Science
8. Woltjer L (1958) A theorem on force-free magnetic fields. Proc Natl Acad Sci 44(6):489–491
9. Taylor JB (1974) Relaxation of toroidal plasma and generation of reverse magnetic fields. Phys Rev Lett 33:1139–1141
10. Seyler CE, Martin MR, Hamlin ND (2018) Helical instability in maglif due to axial flux compression by low-density plasma. Phys Plasmas 25(6):062711
11. Sefkow AB (2016) On the helical instability and efficient stagnation pressure production in thermonuclear magnetized inertial fusion. Bull Am Phys Soc 61(18)
12. Spitzer L (1956) Physics of fully ionized gases. Interscience tracts on physics and astronomy. Interscience Publishers
13. Cvejic M, Mikitchuk D, Doron R, Kroupp E, Maron Y, Velikovich AL, Giuliani JL, Investigation of the properties of the peripheral low-density plasma in z-pinch implosion with and without preembedded axial magnetic field. In: Preparation to Physics of Plasmas

# Chapter 5
# Conclusions

The compression of a plasma with a preembedded axial magnetic field by a Z-pinch implosion and the effects of the compressed field on the plasma implosion and stagnation have been investigated by spectroscopy, 2D imaging, and interferometry. The investigation includes the evolution and spacial distribution of plasma key parameters and magnetic fields.

We report the first measurements of the evolution and distribution of the axial and azimuthal magnetic fields in Z-pinch implosion with preembedded axial $B$-field. The two fields are measured simultaneously. The measurements of $B_z$ and $B_\theta$ employ a novel technique based on the polarization properties of the Zeeman split emission. For the $B_z$ determination we used the $\pi$ and $\sigma$ Zeeman components of the emission from a laser-generated aluminum dopant, and for the $B_\theta$ determination we used the $\sigma^+$ and $\sigma^-$ Zeeman components of the emission from the imploding argon plasma. This first reliable measurement of the compressing and compressed $B$-fields, together with the simultaneous measurement of the plasma radius and discharge current enables us to study the implosion dynamics, pressure balance, current distribution, and $B_z$ distribution.

One of the main (and surprising) result, obtained for implosion with $B_z$, is that the measured $B_\theta$ was much smaller than the expected $B_\theta$, calculated using the total discharge current and the plasma radius. For example, close to stagnation the measured $B_\theta$ is a factor of 5 lower than the expected value. We showed that indeed the presence of $B_z$ causes most of the axial current to flow outside the imploding plasma shell, through a low-density peripheral plasma present at large radii. Moreover, as the implosion progresses, a larger fraction of the axial current flows outside the main plasma, resulting in $B_\theta$ that is almost constant ($\sim 2$ T for $B_{z0} = 0.4$ T) in spite of the implosion. In addition, it was found that, nevertheless the low-density plasma carries a large fraction of the current, it practically doesn't implode on the time scale of the dense plasma implosion. To explain these phenomena we suggested a simplified theory based on the development of a force-free current in the low-density

D. Mikitchuk, *Investigation of the Compression of Magnetized Plasma and Magnetic Flux*, Springer Theses, https://doi.org/10.1007/978-3-030-20855-4_5

plasma. The loss of current and the development of a force-free configuration can also explain different unexpected phenomena observed in other Z-pinch experiments with preembedded axial magnetic field (for more information see [1] and Sect. 4). The detailed data obtained in the present study is currently used for testing MHD codes through the collaboration with plasma groups at the Naval Research Laboratory, the University of Washington, and the University of Rochester.

Simultaneous measurements of $B_z$ and $B_\theta$ at times close to stagnation demonstrate the effect of the plasma inertia on the $B_z$ compression. When the magnetic pressures become equal (i.e. $B_z \approx B_\theta$), the imploding plasma shell still possesses an appreciable inward velocity, further compressing $B_z$ to a value about $3\times$ higher than $B_\theta$. This "overshoot" of $B_z$ compression is expected from theoretical calculations presented in Sects. 1.2.1 and 4.

Knowledge of the compression factor (defined by $B_{z,stagnation}/B_{z0}$), together with the initial and final radii of the plasma, allows for estimating the $B_z$ confinement efficiency relative to a theoretical confinement by an ideal (zero-resistivity) plasma. For the case of $B_{z0} = 0.4$ T, the measurements show $50 \pm 25\%$ confinement efficiency for the studied plasma implosion. This result is important for theoretical studies of the $B$-field diffusion mechanism in plasmas.

The spatially resolved measurements of the axial magnetic field allowed for studying the $B_z$ axial distribution. These measurements revealed an interesting phenomenon of a strong axial gradient of $B_z$, in which the field near the anode ($z \sim 1$ mm) is a factor of $\sim 2$ lower than in the middle of the plasma column ($z \sim 5$ mm). This result shows the existence of a transition region close to the anode, where the axial magnetic field lines embedded in the metal at $t = 0$ remain frozen in the metal, while the magnetic field lines in the plasma near the anode bend as the plasma implodes. This observation should help to understand the effects of the electrodes on the plasma dynamics and on the $B_z$-field evolution and distribution.

Another important part of the research was the investigation of the effects of the preembedded axial magnetic field on the development of magneto-Rayleigh-Taylor instability (MRTI) during the implosion, that are known to bear fundamental importance for the implosion. The analysis of the 2D images and interferograms showed that the presence of an axial magnetic field results in a significant mitigation of the MRTI growth. Furthermore, we observed a rise in the MRTI wavelength for larger $B_{z0}$, indicating the existence of a restoring force that tends to prevent the bending of axial $B$-field lines.

In the present work we also observed another type of instability that appears as axially directed filaments in the plasma (see Fig. 3.20b) which is probably an electro-thermal instability [2, 3]. To investigate this type of instability, the electron densities and temperatures were measured in the generated plasma filaments as well as in the ambient plasma. It was found that the filaments are plasma regions with higher electron density by at least 10–20% relative to the ambient plasma, and that the electron temperature of the filaments is slightly lower (by a few percent) than the temperature of the ambient plasma. These findings lay the basis for a theoretical research on this type of instabilities, which is beyond the scope of the present research.

# References

1. Mikitchuk D, Cvejic M, Doron R, Kroupp E, Stollberg C, Maron Y, Velikovich AL, Ouart ND, Giuliani JL, Mehlhorn TA, Yu EP, Fruchtman A (2019) Effects of a preembedded axial magnetic field on the current distribution in a $Z$-pinch implosion. Phys Rev Lett 122:045001
2. Ryutov DD et al (2000) The physics of fast Z pinches. Rev Mod Phys 72:167
3. Liberman MA, Spielman RB, Toor A, Groot JS (1999) Physics of high-density Z-pinch plasmas. Springer

### References

# Appendix

## A.1 Development of the Force-Free Configuration

Here, we present an analytical solution for the current density and plasma velocity evolution. The analytical solution is made possible in a framework of a simplified model of small $\beta$ plasma in uniform and constant magnetic $\vec{B} = (0, B_\theta, B_z)$, and electric $\vec{E} = (0, 0, E_z)$ fields. Since the $B$-field is constant, the Faraday equation, $\nabla \times \vec{E} = -\frac{\partial \vec{B}}{\partial t}$, is not included. The time scales we derive here are consistent with our conjecture to explain the experimental results. It is emphasized that to obtain a more accurate result for the evolution of $j_z$ and $j_\theta$ distributions in the LDP, one should also consider the self-induced fields and the plasma velocity distribution, which requires numerical modeling. However, it was shown that the inclusion of these effects do not change the tendency of the system to establish a force-free configuration [1, 2] (but might affect the time scale to reach it).

Here, we show the derivation of the current evolution using assumptions described earlier. Since $\beta \ll 1$, pressure terms are omitted from the equations of motion (Eq. A.1) and current evolution (Eq. A.2). In addition, terms of the order $\frac{Zm_e}{M_i}$ and non-linear terms are neglected in Eq. A.2:

$$\frac{d\vec{v}}{dt} = \frac{\vec{j} \times \vec{B}}{\rho} \tag{A.1}$$

$$\frac{d\vec{j}}{dt} = \frac{\nu_{ei}}{\eta}(\vec{E} + \vec{v} \times \vec{B}) - \nu_{ei}\vec{j} - \omega_{ce}\frac{\vec{j} \times \vec{B}}{B} \tag{A.2}$$

Since the plasma has a finite extent in the $r$-direction, with vacuum boundary condition at either side of the plasma, the current density, $j_r$, creates charge separation that in turn generates a radial electrical field (Eq. A.3). In $\hat{\theta}$ and $\hat{z}$ directions no

© Springer Nature Switzerland AG 2019
D. Mikitchuk, *Investigation of the Compression of Magnetized Plasma and Magnetic Flux*, Springer Theses,
https://doi.org/10.1007/978-3-030-20855-4

such charge separation occurs (in the $\hat{\theta}$ direction—due to symmetry, and in the $\hat{z}$ direction—due to the electrodes at the plasma ends).

$$E_r = -\frac{1}{\epsilon_0} \int_0^t j_r(t')dt' \tag{A.3}$$

Differentiating Eq. A.2 gives Eq. A.4 and

$$\frac{d^2\vec{j}}{dt^2} = \frac{\nu_{ei}}{\eta}\left(\frac{d\vec{E}}{dt} + \frac{d\vec{v}}{dt} \times \vec{B}\right) - \nu_{ei}\frac{d\vec{j}}{dt} - \frac{\omega_{ce}}{B}\frac{d\vec{j}}{dt} \times \vec{B} \tag{A.4}$$

Substituting Eqs. A.1 and A.3 into Eq. A.4 gives:

$$\frac{d^2\vec{j}}{dt^2} = \frac{\nu_{ei}}{\eta}\left(\frac{(\vec{j} \times \vec{B}) \times \vec{B}}{\rho} - \frac{\vec{j} \cdot \hat{r}}{\epsilon_0}\right) - \nu_{ei}\frac{d\vec{j}}{dt} - \frac{\omega_{ce}}{B}\frac{d\vec{j}}{dt} \times \vec{B} \tag{A.5}$$

Assuming $j \propto e^{st}$, Eq. A.5 gives:

$$\begin{bmatrix} s^2 + \frac{\nu_{ei}}{\eta\rho}B^2 + \nu_{ei}s + \frac{\nu_{ei}}{\eta\epsilon_0} & \frac{\omega_{ce}}{B}sB_z & -\frac{\omega_{ce}}{B}sB_\theta \\ -\frac{\omega_{ce}}{B}sB_z & s^2 + \frac{\nu_{ei}}{\eta\rho}B_z^2 + \nu_{ei}s & -\frac{\nu_{ei}}{\eta\rho}B_\theta B_z \\ \frac{\omega_{ce}}{B}sB_\theta & -\frac{\nu_{ei}}{\eta\rho}B_\theta B_z & s^2 + \frac{\nu_{ei}}{\eta\rho}B_\theta^2 + \nu_{ei}s \end{bmatrix} \times \begin{bmatrix} j_r \\ j_\theta \\ j_z \end{bmatrix} = 0$$

$$\tag{A.6}$$

Assuming $s \ll \nu_{ei}$, the $s^2$ terms can be neglected. For a non-trivial solution for Eq. A.6, it is required that determinant of the matrix is equal to zero:

$$s\left(\left(1 + \frac{\omega_{ce}^2}{\nu_{ei}^2}\right)s^2 + \left(2\frac{\sigma}{\rho}B^2 + \frac{\sigma}{\epsilon_0}\right)s + \frac{\sigma^2}{\rho^2}B^4 + \frac{\sigma^2}{\rho\epsilon_0}B^2\right) = 0 \tag{A.7}$$

where $\sigma = 1/\eta = \frac{n_e e^2}{\nu_{ei}m_e}$. Using $\omega_{ce} = \frac{eB}{m_e}$, $\omega_{ci} = \frac{ZeB}{A \cdot m_{amu}}$, $\omega_{pl} = \frac{n_e e^2}{m_e \epsilon_0}$, Eq. A.7 becomes

$$s\left(\left(1 + \frac{\omega_{ce}^2}{\nu_{ei}^2}\right)s^2 + \left(2\frac{\omega_{ce}\omega_{ci}}{\nu_{ei}} + \frac{\omega_{pl}^2}{\nu_{ei}}\right)s + \frac{\omega_{ce}^2\omega_{ci}^2}{\nu_{ei}^2} + \frac{\omega_{ce}\omega_{ci}\omega_{pl}^2}{\nu_{ei}^2}\right) = 0 \tag{A.8}$$

The roots are:

$$s_0 = 0,$$

$$s_{1,2} = \frac{-\left(2\frac{\omega_{ce}\omega_{ci}}{\nu_{ei}} + \frac{\omega_{pl}^2}{\nu_{ei}}\right) \pm \sqrt{\frac{\omega_{pl}^4}{\nu_{ei}^2} - 4\frac{\omega_{ce}^4\omega_{ci}^2}{\nu_{ei}^4} - 4\frac{\omega_{ce}^3\omega_{ci}\omega_{pl}^2}{\nu_{ei}^4}}}{2\left(1 + \frac{\omega_{ce}^2}{\nu_{ei}^2}\right)} \tag{A.9}$$

For the plasma and magnetic field parameters relevant to our experiments ($10^{16} <$ $n_e < 10^{17}$ cm$^{-3}$, $T_e \sim 5$ eV, $B \sim 1$ T): $\omega_{pl} \gg \omega_{ce} \gg \omega_{ci}$. The roots can then be approximated by:

$$s_1 = -\frac{\omega_{ce}\omega_{ci}}{\nu_{ei}},$$

$$s_2 = \frac{\omega_{ce}\omega_{ci}}{\nu_{ei}} \frac{1 - \frac{\omega_{ce}^2}{\nu_{ei}^2} + \frac{\omega_{pl}^2}{\omega_{ce}\omega_{ci}}}{1 + \frac{\omega_{ce}^2}{\nu_{ei}^2}}$$

(A.10)

$\vec{j}$ that corresponds to root $s_0$ is a force-free configuration geometry $(0, j_\theta, j_\theta \frac{B_z}{B_\theta})$. Besides $s_0$, the time scale of $1/s_1$ is the slowest, and determines the time scale for establishing the force-free configuration. For our LDP parameters $s_1 \approx 10 - 30$ ns, which is more than an order of magnitude smaller than the implosion time. We note that $s_1$ is the only root (besides $s_0$) consistent with the assumption of $s \ll \nu_{ei}$, i.e the derivation provides an accurate result for $s_1$, but it is not valid for $s_2$. However, since we are interested in the slowest time scale, this derivation provides an accurate result.

We also note that retaining the $s^2$ term in Eq. A.6 and solving numerically the determinant polynomial give the following roots when $\omega_{pl} \gg \omega_{ce}$:

$$\tilde{s}_0 = 0$$

$$\tilde{s}_1 \approx -\frac{\omega_{ce}\omega_{ci}}{\nu_{ei}};$$

$$\tilde{s}_{2,3} \approx -\nu_{ei};$$

$$\tilde{s}_{4,5} \approx -\frac{\nu_{ei}}{2} \pm i\omega_{pl};$$

(A.11)

As expected $s_1$ is not changed.

After finding the roots $s$, the general solution for $\vec{j}(t)$ can be written in the following form:

$$\vec{j} = A_0 \hat{V}_0 e^{s_0 t} + A_1 \hat{V}_1 e^{s_1 t} + A_2 \hat{V}_2 e^{s_2 t} + A_3 \hat{V}_3 e^{s_3 t} + A_4 \hat{V}_4 e^{s_4 t} + A_5 \hat{V}_5 e^{s_5 t} \quad \text{(A.12)}$$

where $\hat{V}_0$, $\hat{V}_1$, $\hat{V}_2$, $\hat{V}_3$, $\hat{V}_4$, and $\hat{V}_5$ are the eigenvectors of the matrix in Eq. A.6 corresponding to the roots $s_0$, $s_1$, $s_2$, $s_3$, $s_4$, and $s_5$, respectively. These eigenvectors are:

$$\hat{V}_0 = \begin{bmatrix} 0 \\ \frac{1}{\sqrt{2}} \\ \frac{1}{\sqrt{2}} \end{bmatrix} \quad \hat{V}_1 = \begin{bmatrix} 0 \\ \frac{-1}{\sqrt{2}} \\ \frac{1}{\sqrt{2}} \end{bmatrix} \quad \hat{V}_2 = \begin{bmatrix} 0 \\ \frac{-1}{\sqrt{2}} \\ \frac{1}{\sqrt{2}} \end{bmatrix} \quad \hat{V}_3 = \begin{bmatrix} 0 \\ \frac{-1}{\sqrt{2}} \\ \frac{1}{\sqrt{2}} \end{bmatrix} \quad \hat{V}_4 = \begin{bmatrix} 1 \\ 0 \\ 0 \end{bmatrix} \quad \hat{V}_5 = \begin{bmatrix} 1 \\ 0 \\ 0 \end{bmatrix}$$

(A.13)

The coefficients $A_0, A_1, A_2, A_3, A_4$, and $A_5$ are found from the initial conditions $\vec{j}(t = 0)$ and $\frac{d\vec{j}}{dt}(t = 0)$. In practice the known initial parameters are $\vec{j}(t = 0)$ and $\vec{v}(t = 0)$, then $\frac{d\vec{j}}{dt}(t=0)$ can be found using Eq. A.2. The evolution of plasma velocity is then calculated by the time integration of Eq. A.1 and using the solution for the current density evolution (Eq. A.12).

As can be seen from Eq. A.11 the real parts of the roots are either negative or zero, which means that the current evolution given by the Eq. A.12 converges asymptotically to a steady state solution. The steady state solution is defined by the $A_0$ coefficient and can be found by taking $\frac{d\vec{v}}{dt}$ and $\frac{d\vec{j}}{dt}$ in Eqs. A.1 and A.2 to be zero:

$$\vec{j}_{steady} = \begin{bmatrix} 0 \\ \frac{E_z}{\eta} \frac{B_z B_\theta}{B_z^2 + B_\theta^2} \\ \frac{E_z}{\eta} \frac{B_z^2}{B_z^2 + B_\theta^2} \end{bmatrix} \qquad \vec{v}_{steady} = \begin{bmatrix} \frac{E_z B_\theta}{B_z^2 + B_\theta^2} \\ 0 \\ 0 \end{bmatrix} \qquad (A.14)$$

## A.2   Electron Motion in Constant Electric and Magnetic Fields Without Collisions: General Solution

Like in Appendix A.3, also here the cylindrical geometry of the experiment is approximated by a planar geometry (see Appendix A.3 for more details).

The equation of motion of an electron in collisionless plasma subjected to constant magnetic $\vec{B} = (0, B_y, B_z)$ and electric $\vec{E} = (0, 0, E_z)$ fields is:

$$m_e \frac{dv_x}{dt} = -ev_y B_z + ev_z B_y \qquad (A.15a)$$

$$m_e \frac{dv_y}{dt} = ev_x B_z \qquad (A.15b)$$

$$m_e \frac{dv_z}{dt} = -eE_z - ev_x B_y \qquad (A.15c)$$

The time derivative of both sides of Eq. A.15a and using Eqs. A.15b and A.15c gives:

$$\frac{d^2 v_x}{dt^2} = -\frac{e}{m_e} \frac{dv_y}{dt} B_z + \frac{e}{m_e} \frac{dv_z}{dt} B_y = -\frac{e^2}{m_e^2}(B_y^2 + B_z^2)v_x - \frac{e^2}{m_e^2} E_z B_y \qquad (A.16)$$

The solution of Eq. A.16 is:

$$v_x = A \sin\left(\frac{e}{m}\sqrt{B_y^2 + B_z^2}\,t + \phi\right) - E_z \frac{B_y}{B_y^2 + B_z^2} \qquad (A.17)$$

Equations A.15b and A.15c can be solved after substitution of the solution for $v_x$ (see Eq. A.17):

$$v_y = -A \frac{B_z}{\sqrt{B_y^2 + B_z^2}} \cos\left(\frac{e}{m}\sqrt{B_y^2 + B_z^2}\, t + \phi\right) - \frac{eE_z}{m_e} \frac{B_y B_z}{B_y^2 + B_z^2} t + C \quad (A.18)$$

$$v_z = A \frac{B_y}{\sqrt{B_y^2 + B_z^2}} \cos\left(\frac{e}{m}\sqrt{B_y^2 + B_z^2}\, t + \phi\right) - \frac{eE_z}{m_e} \frac{B_z^2}{B_y^2 + B_z^2} t + D \quad (A.19)$$

By substitution of $v_x$, $v_y$, and $v_z$ solutions into Eq. A.15a, the relation between the integration constants $C$ and $D$ is found: $\frac{C}{D} = \frac{B_y}{B_z}$.

Therefore, the set of solutions of Eqs. A.15a–A.15c is:

$$v_x = A \sin\left(\frac{e}{m}\sqrt{B_y^2 + B_z^2}\, t + \phi\right) - E_z \frac{B_y}{B_y^2 + B_z^2} \quad (A.20a)$$

$$v_y = -A \frac{B_z}{\sqrt{B_y^2 + B_z^2}} \cos\left(\frac{e}{m}\sqrt{B_y^2 + B_z^2}\, t + \phi\right) - \frac{eE_z}{m_e} \frac{B_y B_z}{B_y^2 + B_z^2} t + C \quad (A.20b)$$

$$v_z = A \frac{B_y}{\sqrt{B_y^2 + B_z^2}} \cos\left(\frac{e}{m}\sqrt{B_y^2 + B_z^2}\, t + \phi\right) - \frac{eE_z}{m_e} \frac{B_z^2}{B_y^2 + B_z^2} t + C \frac{B_z}{B_y} \quad (A.20c)$$

The constants $A$, $\phi$, and $C$ are determined from the initial velocity of the electron.

For example, in the case of an electron starting from rest, $\vec{v} = 0$, the constants are: $A = E_z \frac{B_y}{B_y^2 + B_z^2}$, $\phi = \frac{\pi}{2}$, $C = 0$ and the full solution is:

$$v_x = E_z \frac{B_y}{B_y^2 + B_z^2}\left(\cos\left(\frac{e}{m}\sqrt{B_y^2 + B_z^2}\, t\right) - 1\right) = -2E_z \frac{B_y}{B_y^2 + B_z^2} \sin^2\left(\frac{e}{2m}\sqrt{B_y^2 + B_z^2}\, t\right) \quad (A.21a)$$

$$v_y = E_z \frac{B_y B_z}{(B_y^2 + B_z^2)^{3/2}} \sin\left(\frac{e}{m}\sqrt{B_y^2 + B_z^2}\, t\right) - \frac{eE_z}{m_e} \frac{B_y B_z}{B_y^2 + B_z^2} t \quad (A.21b)$$

$$v_z = -E_z \frac{B_y^2}{(B_y^2 + B_z^2)^{3/2}} \sin\left(\frac{e}{m}\sqrt{B_y^2 + B_z^2}\, t\right) - \frac{eE_z}{m_e} \frac{B_z^2}{B_y^2 + B_z^2} t \quad (A.21c)$$

By time averaging the Eqs. A.21a–A.21c we obtain the guiding center velocity:

$$v_x = -E_z \frac{B_y}{B_y^2 + B_z^2} \quad (A.22a)$$

$$v_y = -B_y \frac{eE_z}{m_e} \frac{B_z}{B_y^2 + B_z^2} t \quad (A.22b)$$

$$v_z = -B_z \frac{eE_z}{m_e} \frac{B_z}{B_y^2 + B_z^2} t \quad (A.22c)$$

From expressions (A.22b) and (A.22c) it is seen that the projected electron velocity on the $yz$ plane is in the direction of the magnetic field.

It is important to note that the calculations presented here are not self-consistent since the magnetic field generated by the considered electrons is not taken into account. The electrons are in the externally generated electric and magnetic fields, and the magnetic field they produce is small in comparison to the external fields. This case is applicable for the initial stage of the implosion with $B_{z0} = 0.4$ T, when a small part of the total current is flowing through the low-density plasma, such that $B_\theta$ is generated externally by the main current flowing in the imploding dense plasma and $B_z$ is generated by the Helmholtz coils.

## A.3  Electron Motion in Constant Electric and Magnetic Fields with Collisions: Steady State Solution

In order to obtain an analytical solution to the problem of electron motion in constant electric and magnetic fields we approximate the cylindrical geometry of the experimental setup by a planar geometry, where $B_\theta$ is replaced by $B_y$ and the $r$-coordinate of the cylindrical geometry is in the $x$ direction of the planar geometry. This approximation is valid to describe electron dynamics if the plasma radius is much bigger than the electron gyro-radius.

The equation of motion of electrons in collisional plasma subjected to constant magnetic $\vec{B} = (0, B_y, B_z)$ and electric $\vec{E} = (0, 0, E_z)$ fields is:

$$m_e \frac{dv_x}{dt} = -e v_y B_z + e v_z B_y - m_e v_x \nu_{ei} \qquad (A.23a)$$

$$m_e \frac{dv_y}{dt} = e v_x B_z - m_e v_y \nu_{ei} \qquad (A.23b)$$

$$m_e \frac{dv_z}{dt} = -e E_z - e v_x B_y - m_e v_z \nu_{ei} \qquad (A.23c)$$

where $e$ is the elementary charge, $\nu_{ei}$ is electron-ion collision frequency and $\vec{v} = (v_x, v_y, v_z)$ is the electron velocity.

In steady state $\frac{d\vec{v}}{dt} = 0$, therefore Eq. A.23 become:

$$0 = -e v_y B_z + e v_z B_y - m_e v_x \nu_{ei} \qquad (A.24a)$$

$$0 = e v_x B_z - m_e v_y \nu_{ei} \quad \Rightarrow \quad v_y = \frac{e v_x B_z}{m_e \nu_{ei}} = \frac{\omega_{ce}^{(z)}}{\nu_{ei}} v_x \qquad (A.24b)$$

$$0 = -e E_z - e v_x B_y - m_e v_z \nu_{ei} \quad \Rightarrow \quad v_z = -\frac{e E_z + e v_x B_y}{m_e \nu_{ei}} = -v_{drift} - \frac{\omega_{ce}^{(y)}}{\nu_{ei}} v_x \qquad (A.24c)$$

where $\omega_{ce}^{(y)} \equiv \frac{eB_y}{m_e}$, $\omega_{ce}^{(z)} \equiv \frac{eB_z}{m_e}$, and $v_{drift} \equiv \frac{eE_z}{m_e \nu_{ei}}$. $v_{drift}$ is the electron drift velocity without magnetic field or in collisional-dominated plasma (i.e. $\nu_{ei} \gg \omega_{ce}$).

Substituting $v_y = \frac{\omega_{ce}^{(z)}}{\nu_{ei}} v_x$ and $v_z = -v_{drift} - \frac{\omega_{ce}^{(y)}}{\nu_{ei}} v_x$ into Eq. A.24a we obtain:

$$v_x = -\frac{\frac{\omega_{ce}^{(y)}}{\nu_{ei}} v_{drift}}{1 + \left( \frac{(\omega_{ce}^{(y)})^2 + (\omega_{ce}^{(z)})^2}{\nu_{ei}^2} \right)} \tag{A.25}$$

By substitution of $v_x$ solution (Eq. A.25) into Eqs. A.24b and A.24c the expressions for $v_y$ and $v_z$ are found:

$$v_y = -\frac{\frac{\omega_{ce}^{(y)} \omega_{ce}^{(z)}}{\nu_{ei}^2} v_{drift}}{1 + \left( \frac{(\omega_{ce}^{(y)})^2 + (\omega_{ce}^{(z)})^2}{\nu_{ei}^2} \right)} \tag{A.26}$$

$$v_z = -\frac{\left( 1 + \left( \frac{\omega_{ce}^{(z)}}{\nu_{ei}} \right)^2 \right) v_{drift}}{1 + \left( \frac{(\omega_{ce}^{(y)})^2 + (\omega_{ce}^{(z)})^2}{\nu_{ei}^2} \right)} \tag{A.27}$$

For implosion with $B_{z0} = 0$ (i.e. $\omega_{ce}^{(z)} = 0$) and using Eq. A.27 we see that if $\omega_{ce}^{(y)} \gg \nu_{ei}$ (equivalent to small $n_e$ and large $B_y$) then $v_z \approx \left( \frac{\nu_{ei}}{\omega_{ce}^{(y)}} \right)^2 v_{drift} \ll v_{drift}$. Such conditions might be relevant for the electrons of the low-density plasma in implosions with $B_{z0} = 0$. For implosion with $B_{z0} > 0$ and $\nu_{ei} \ll \omega_{ce}^{(y)}, \omega_{ce}^{(z)}$, Eq. A.27 gives: $v_z \approx -v_{drift} \frac{B_z^2}{B_y^2 + B_z^2}$. Such conditions might be relevant for the electrons of the low-density plasma in implosions with $B_{z0} = 0.4$ T at the early times of the current flow in the low-density plasma (see Chap. 4).

# References

1. Woltjer L (1958) A theorem on force-free magnetic fields. Proceedings of the National Academy of Sciences 44(6):489–491
2. Taylor JB (1974) Relaxation of toroidal plasma and generation of reverse magnetic fields. Phys. Rev. Lett. 33:1139–1141

Printed in the United States
By Bookmasters